Protein Methods

Protein Methods

Second Edition

DANIEL M. BOLLAG
Merck Research Laboratories
West Point, Pennsylvania

MICHAEL D. ROZYCKI
Department of Chemistry
The Henry H. Hoyt Laboratory
Princeton University
Princeton, New Jersey

STUART J. EDELSTEIN
Department of Biochemistry
University of Geneva
Geneva, Switzerland

 **WILEY-LISS**

A JOHN WILEY & SONS, INC., PUBLICATION

New York • Chichester • Brisbane • Toronto • Singapore

Address all Inquiries to the Publisher
Wiley-Liss, Inc., 605 Third Avenue, New York, NY 10158-0012

Copyright © 1996 Wiley-Liss, Inc.

Printed in the United States of America.

The text of this book is printed on acid-free paper.

While the authors, editors, and publisher believe that drug selection and dosage and the specification and usage of equipment and devices, as set forth in this book are in accord with current recommendations and practice at the time of publication, they accept no legal responsibility for any errors or omissions, and make no warranty, express or implied, with respect to material contained herein. In view of ongoing research, equipment modifications, changes in governmental regulations and the constant flow of information relating to drug therapy, drug reactions, and the use of equipment and devices, the reader is urged to review and evaluate the information provided in the package insert or instructions for each drug, piece of equipment, or device for, among other things, any changes in the instructions or indication of dosage or usage and for added warnings and precautions.

Library of Congress Cataloging in Publication Data

Bollag, Daniel M.
    Protein methods / Daniel M. Bollag, Michael D. Rozycki.
Stuart J. Edelstein. — 2nd ed.
      p.   cm.
    Includes bibliographical references and index.
    ISBN 0-471-11837-0 (cloth : alk. paper)
    1. Proteins—Purification. 2. Proteins—Research—Methodology.
  I. Edelstein, Stuart J. II. Rozycki, Michael D. III. Title.
QP551.E23 1996
574.19'245—dc20   96-14083

10 9 8 7 6 5 4 3 2 1

Contents

Preface   xiii

1. Preparation for Protein Isolation   1

2. Protein Extraction and Solubilization   27

3. Protein Concentration Determination   57

4. Concentrating Protein Solutions   83

5. Gel Electrophoresis Under Denaturing Conditions   107

6. Gel Electrophoresis Under Nondenaturing Conditions   155

7. Isoelectric Focusing and Two-Dimensional Gel Electrophoresis   173

8. Immunoblotting   195

9. Ion Exchange Chromatography   231

10. Gel Filtration Chromatography   271

11. Affinity Chromatography   301

12. Hanging Drop Crystallization   353

Appendix 1: Molecular Weights of Commonly Used Chemicals   389

Appendix 2: Molecular Weights and Isoelectric Points of Selected Proteins   393

Appendix 3: Ammonium Sulfate Precipitation Table   394

Appendix 4: Spectrophotometer Linearity   397

Appendix 5: Suppliers and Addresses   399

Index   405

Preface xiii

PART ONE

1. Preparation for Protein Isolation 1

   I. Introduction 2

   II. Buffers 3
      A. Buffer Characteristics 3
      B. Preparation of Buffers 5
      C. Concentration Effects of Buffer on pH 8
      D. Limitations of Certain Buffers 8
      E. Preventing Buffer Contamination 10
      F. Water Purity 10

   III. Salts, Metal Ions, and Chelators 11
      A. Ionic Strength 11
      B. Divalent Cations 11
      C. Chelators 11

   IV. Reducing Agents 12
      A. General Considerations 12
      B. Specific Recommendations 12

   V. Detergents 13
      A. Introduction 13
      B. Classes of Detergents 14
      C. Protocol for Membrane Protein Solubilization 18

   VI. Protein Environment 20
      A. Surface Effects 20
      B. Temperature 20
      C. Storage 21

   VII. Protease Inhibitors 23
      A. Common Inhibitors 23
      B. A Sample Broad Range Protease Inhibitor
         Cocktail 24

   VIII. References 25

2. Protein Extraction and Solubilization 27

   I. Introduction 29

   II. Protein Extraction 30
      A. Homogenization 32
         1. General Considerations 32
         2. Specific Steps 33
      B. Sonication 34
         1. General Considerations 34
         2. Specific Steps 35

C. French Pressure Cell 36
    1. General Considerations 36
    2. Specific Steps 37
D. Grinding With Alumina or Sand 38
    1. General Considerations 38
    2. Specific Steps 38
E. Glass Bead Vortexing 39
    1. General Considerations 39
    2. Specific Steps 40
F. Enzymatic Treatments 41
    1. General Considerations 41
    2. Specific Steps 41
G. Other Lysis Methods 42
    1. Detergent Lysis 42
    2. Organic Solvent Lysis 43
    3. Osmotic Shock Lysis 43
    4. Freeze-Thaw Lysis 43

III. Solubilization of Protein From Inclusion Bodies 44
   A. Solubilization of Inclusion Body Protein 45
       1. General Considerations 45
       2. Specific Steps 46
   B. Prevention of Inclusion Body Formation 50

IV. Suppliers 52

V. References 53

3. **Protein Concentration Determination** **57**

I. Absorbance at 280 nm (A$_{280}$) 58
    A. Summary 58
    B. Equipment 58
    C. Reagents 58
    D. Protocol 59
    E. Comments 59

II. Bradford Assay 62
     A. Summary 62
     B. Equipment 62
     C. Reagents 62
     D. Protocol 63
     E. Comments 67

III. Lowry Assay 68
      A. Summary 68
      B. Equipment 68
      C. Reagents 68
      D. Protocol 69
      E. Comments 71

    IV. Bicinchoninic Acid (BCA) Assay 72
        A. Summary 72
        B. Equipment 72
        C. Reagents 72
        D. Protocol 73
        E. Comments 74

     V. Briefly Noted: Dot Filter Binding Assay 75
        A. Summary 75
        B. Protocol 75

    VI. Interference Table 76

   VII. Suppliers 79

  VIII. References 80

4. Concentrating Protein Solutions 83

      I. Analytical Methods 84
        A. Trichloroacetic Acid (TCA) Precipitation 84
        B. Acetone Precipitation 86
        C. Immunoprecipitation 87

     II. Preparative Methods 91
        A. Ammonium Sulfate Precipitation ("Salting Out") ... 91
        B. Organic Solvent Precipitation 94
        C. Poly(ethylene glycol) (PEG) Precipitation 96
        D. Ultrafiltration 97
        E. Dialysis 100
        F. Briefly Noted: Ion Exchange Chromatography and
           Lyophilization 103

    III. Suppliers 104

    IV. References 105

5. Gel Electrophoresis Under Denaturing Conditions 107

      I. SDS-Polyacrylamide Gel Electrophoresis (Linear
        Slab Gel) 108
        A. Introduction 108
        B. Equipment 109
        C. Pouring a Gel 110
        D. Preparing and Loading Samples 121
        E. Running a Gel 124
        F. Staining a Gel with Coomassie Blue 126
        G. Silver Staining a Gel 129
        H. Drying a Gel 135
        I. General Discussion 136
        J. Safety Notes 139

II.   Gradient Gels 140
      A. Introduction 140
      B. Equipment 140
      C. Preparing Gels 140

III. SDS-Urea Gels 143
      A. Introduction 143
      B. Preparing Gels 143

IV.   Other Methods 145
      A. Detection of Radiolabeled Samples 145
      B. Molecular Weight Determination 146
      C. Protein Quantitation (Densitometry) 148
      D. Eluting Protein Bands Following
          Electrophoresis 149

V.   Suppliers 150

VI.   References 151

6. Gel Electrophoresis Under Nondenaturing Conditions 155

I.   Introduction 156

II.   Discontinuous Nondenaturing Gel Electrophoresis . . . 157
      A. Introduction 157
      B. Equipment 157
      C. Preparing the Gel 157
      D. Sample Preparation 164
      E. Running the Gel 166
      F. Staining the Gel 167
      G. Variation: Continuous Nondenaturing Gel
          Electrophoresis 168

III. Related Methods 169
      A. Determining Protein Molecular Weight 169
      B. Determining Enzyme Activity After
          Electrophoresis 171

IV.   References 172

7. Isoelectric Focusing and Two-Dimensional Gel
    Electrophoresis 173

I.   Isoelectric Focusing (IEF) 174
      A. Introduction 174
      B. Equipment 175
      C. Preparing Focusing Gel 176
      D. Sample Preparation and Loading 180
      E. Running Isoelectric Focusing 181
      F. Post-Focusing Procedures 182

x   Contents

      G. Modifications for a Native Isoelectric Focusing
         Gel 183
      H. Discussion 184

   II. Two-Dimensional Gel Electrophoresis 188
      A. Introduction 188
      B. Equipment 188
      C. Protocols 188
      D. Discussion 191

   III. Suppliers 192

   IV. References 193

8. **Immunoblotting** **195**

   I. Introduction 196

   II. Performing an Immunoblot 199
      A. Equipment 199
      B. Reagents 200
      C. Protocols 201
      D. Protocol Modifications for Other Detection
         Methods 213
      E. Staining for Total Protein 216
      F. Erasing Immunoblots 218

   III. Discussion 219
      A. Membrane Storage 219
      B. Transfer Anomalies 219
      C. References for Other Uses of Immunoblots 219

   IV. Suppliers 222

   V. References 223

PART TWO

**PURIFICATION AND CRYSTALLIZATION OF
   PROTEINS** **229**

9. **Ion Exchange Chromatography** **231**

   I. Protein Purification 232
      A. Introduction 232
      B. Methods 243

   II. Concentrating a Protein Solution 266

   III. Batch Chromatography 267

   IV. Suppliers 268

   V. References 269

10. Gel Filtration Chromatography **271**

    I. Protein Purification 272
      A. Introduction 272
      B. Methods 281

    II. Exchanging the Buffer of a Protein 296

    III. Suppliers 298

    IV. References 299

11. Affinity Chromatography **301**

    I. Introduction 301

    II. Preparing an Affinity Column 303
      A. Ligand Immobilization on Cyanogen
         Bromide-Activated Sepharose 307
      B. Nonspecific Elution Strategies 310

    III. Specialized Techniques 313
      A. Antibodies 313
      B. Nucleic Acid Affinity Chromatography 326
      C. Lectin 333
      D. Dye Ligand Chromatography 338
      E. Immobilized Metal Affinity Chromatography 343
      F. Hydrophobic Interaction Chromatography 346

    IV. References 349

12. Hanging Drop Crystallization **353**

    I. Introduction 354

    II. Performing a Hanging Drop Crystallization 356
      A. Crystallization Principles 356
      B. Procedure 361

    III. Designing an Optimization Strategy 376
      A. Fine-tuning the Sparse Matrix Conditions 376
      B. Expanding the Initial Screen 384
      C. Still No Crystals? 384

    IV. Suppliers 385

    V. References 386

**Appendix 1: Molecular Weights of Commonly Used
Chemicals** **389**

**Appendix 2: Molecular Weights and Isoelectric Points of
Selected Proteins** **393**

Appendix 3: Ammonium Sulfate Precipitation Table **394**

xii   Contents

Appendix 4: Spectrophotometer Linearity **397**
Appendix 5: Suppliers and Addresses **399**
Index **405**

Preface to the Second Edition

The motivation for writing the first edition of this book was to describe techniques of general applicability to proteins. The positive response to *Protein Methods* encouraged us to prepare an updated version and to expand the scope to include procedures that require considering the specific characteristics of the protein under investigation. Hence, in addition to the eight chapters that appeared in the first edition, we have added three chapters on methods of purification that must be tailored to the properties of individual proteins. In view of the burgeoning activity in the determination of protein atomic structure by X-ray crystallography, we also decided to include a chapter on crystallization. In addition, since in many instances, proteins are now purified following expression in bacteria, a new section was added to Chapter 2 concerning the isolation of proteins from inclusion bodies.

In order to tackle the widened range of topics, the authors of the first edition, Daniel M. Bollag and Stuart J. Edelstein, invited Michael D. Rozycki to participate in writing the second edition in order to benefit from his knowledge of protein methods in general and in particular from his extensive experience in protein crystallography. We were fortunate to have comments on the crystallization chapter from Nancy Vogelaar, Rebecca Page, Greg Bowman, Dale Jefferson, and Donatella Pascolini. From Wiley, the encouragement of Peter Brown to begin work on a second edition was decisive and in later stages the help of Susan King and Colette Bean was very much appreciated. The contributions of all those acknowledged in the first edition continue to be appreciated and we reiterate our thanks to friends and colleagues who contributed in various ways. Most of all, we are grateful to our wives, Wendy Miller, Teresa Twomey, and Lynn Edelstein, for their constant support during this effort to expand the scope of *Protein Methods*.

Preface to the First Edition

The revolution in genetic engineering has benefitted greatly from the fact that the behavior of DNA fragments created by restriction enzymes is largely independent of the fragments' precise compositions. Many properties of these fragments therefore depend essentially on their size only. This feature of DNA has permitted various generalized methods to be developed and summarized in extremely useful laboratory manuals, notably *Molecular Cloning* by Maniatis, Fritsch, and Sambrook.

Proteins, from many points of view, have more complicated personalities than DNA fragments. As a result, it has been difficult to design manuals for laboratory methods that can be applied to proteins in general. However, for routine methods concerning the elementary operations of extraction and concentration determination, as well as for certain widely used techniques involving gel electrophoresis, the distinctive properties of different proteins are not of primary importance. Hence, these methods can be applied directly to virtually all proteins.

Our goal in preparing this book has been to assemble the most general methods for protein research and to present them in practical detail in order to provide investigators with all of the information needed to perform these procedures in the laboratory. In addition, we have organized the material in the form of standardized laboratory protocols to facilitate its utilization. This orientation has led us to delve only minimally into the theory of the methods or into the basic concepts of protein biochemistry, but other sources are available for these topics.

Several guidelines have served us in designing the nature of the topics and the manner in which they are presented. The most common theme of protein analysis at the current time is gel electrophoresis, and this serves as the heart of *Protein Methods* (Chapters 5, 6, and 7). The extension of gel electrophoresis involving electroblotting and immunochemical detection of electroblotted proteins has led to some of the most impressive advances in protein analysis in recent years, and these techniques are treated in Chapter 8. Protein analysis by electrophoretic methods requires that the protein sample be extracted from its native cellular environment (Chapter 2), that the *in vitro* protein concentration be estimated (Chapter 3), and sometimes that the sample be concentrated prior to analysis (Chapter 4). Some fundamental principles for handling of proteins are outlined in Chapter 1.

In many cases, the descriptions presented are derived from procedures that have been used and refined in our own laboratory. However, their origins are multiple and are not always accurately traceable. We sincerely apologize to any uncited investigators who may be responsible for key developments. Overall, we have attempted to provide sufficient references to the original literature to enable each researcher to expand his or her knowledge of the subject and to develop specialized refinements.

We have submitted the material presented here to various colleagues for their critical evaluation, and we particularly wish to express our appreciation to Christine Tachibana and Gideon Bollag who provided valuable comments during the development of the manuscript, as well as to Clement Bordier for his guidance on the sections concerning detergents. We thank Isabelle Tornare and Ann-Marie Paunier Doret in our laboratory for their participation in many of the developments that led to the final protocols. The skillful contributions of F. Pillonel in the preparation of the figures are gratefully acknowledged. For the weak points or errors that may remain, we take full responsibility.

A methods book of this type can always be improved by feedback from users, and we are eager to hear from those who use this book and wish to share with us their comments, criticisms, and suggestions for additions or deletions.

We are most appreciative of the people at the Wiley-Liss Division of John Wiley & Sons, Inc. who helped with this project, starting with Peter Brown, with whom it was conceived, and including Eric Swanson, Eileen Cudlipp, Frederick Siebenmann, and Sonny Fritz. Gary Giulian, Henry Reeves, and Kathleen Dannelly kindly supplied figures used in Chapters 5 and 7.

We also thank our family, friends, and colleagues upon whom we imposed for reading various sections, providing advice, and tolerating our preoccupations during the preparation of this book.

Chapter 1

Preparation for Protein Isolation

I. Introduction

II. Buffers
  A. Buffer Characteristics
  B. Preparation of Buffers
  C. Concentration Effects of Buffer on pH
  D. Limitations of Certain Buffers
  E. Preventing Buffer Contamination
  F. Water Purity

III. Salts, Metal Ions, and Chelators
  A. Ionic Strength
  B. Divalent Cations
  C. Chelators

IV. Reducing Agents
  A. General Considerations
  B. Specific Recommendations

V. Detergents
  A. Introduction
  B. Classes of Detergents
  C. Protocol for Membrane Protein Solubilization

VI. Protein Environment
  A. Surface Effects
  B. Temperature
  C. Storage

VII. Protease Inhibitors
  A. Common Inhibitors
  B. A Sample Broad Range Protease Inhibitor Cocktail

VIII. References

I. Introduction

This book is devoted to laboratory techniques for the analysis and separation of proteins. Proteins are an extremely heterogeneous class of biological macromolecules. They are often unstable when not in their native environment, which in itself varies considerably among cell compartments and extracellular fluids. Of the many types of proteins, we can distinguish between those that are soluble or membrane-bound, those with catalytic or purely structural roles, and those with various post-translational modifications.

Each protein may have specific requirements once it is extracted from its normal biological milieu. If these requirements are not satisfied, the protein can rapidly lose its ability to carry out specific functions, and an already limited lifetime may be drastically reduced. Thus, determination of these requirements has often been a major hurdle in protein characterization. In some cases, the difficulty has been to stabilize the protein against external proteolysis, while in other cases the problem has been to maintain ligand-binding or enzymatic activity. Solutions to these problems are highly individual. Nonetheless, some fundamental parameters must be considered by anyone studying proteins. In this chapter, we discuss a number of these parameters and attempt to provide general guidelines or sources of information for laboratory work with proteins.

II. Buffers

A. Buffer Characteristics

- A buffer is defined as a mixture of an acid and its conjugate base which can reduce changes in solution pH when acid or alkali are added. The selection of an appropriate buffer is important in order to maintain a protein at the desired pH and to ensure reproducible experimental results. A rudimentary description of key concepts behind buffering, such as pH and pK_a, can be found in the Calbiochem "Buffers" booklet and in Stryer (1988, pp. 41-42).

- There are eight important characteristics to consider when selecting a buffer (adapted from Scopes, 1982):
    1. pK_a value (see Table 1.1)
    2. pK_a variation with temperature
    3. pK_a variation upon dilution
    4. Solubility
    5. Interaction with other components (such as metal ions and enzymes)
    6. Expense
    7. UV absorbance
    8. Permeability through biological membranes

- Some General Observations

    1. Ideally, different buffers with a similar pK_a should be tested to determine whether there are undesired interactions between a certain buffer and the protein under investigation (Blanchard, 1984).

    2. Once a buffer is chosen, it is best to work at the lowest reasonable concentration to avoid nonspecific ionic strength effects. A 50 mM buffer is a good starting point.

3. The useful buffering range diminishes significantly beyond 1 pH unit on either side of the pK_a. Note that many enzymes are irreversibly denatured at extreme pH values (Tipton and Dixon, 1979).

4. The physiological pH in most animal cells is 7.0 - 7.5 at 37°C. Due to the effect of temperature, this value rises to close to 8.0 near 0°C (Scopes, 1982).

5. The buffer of choice also depends on the methods employed:
   - For gel filtration chromatography, almost any buffer can be chosen that is compatible with the protein of interest.
   - For anion exchange chromatography, cationic buffers such as Tris are preferred.
   - For cation exchange or hydroxyapatite chromatography, anionic buffers such as phosphate are preferred (Blanchard, 1984).

6. 'Good's' buffers (for example, MES, PIPES, MOPS) were developed by Good and colleagues (1966) to be biologically inert, to have low UV absorbance, and to be minimally affected by temperature or ionic strength.

7. Buffer mixtures with wide buffering ranges at constant ionic strength are described by Ellis and Morrison (1982).

8. A description of buffers and cryosolvents for low temperature conditions is found in Fink and Geeves (1979).

9. All chemical products should be reagent grade or higher.

B. Preparation of Buffers

- In principle, the pH of a solution can be adjusted directly at the temperature at which the buffer is to be used. However, this requires that the pH electrode be standardized at the working temperature, often 4°C, while in practice, buffer is usually prepared at room temperature and the pH adjusted so that it will be correct after the solution is brought to the desired temperature. It should be noted that temperature effects on buffer pH may be large. A notable example is Tris, which has a pK_a that changes from 8.06 at 25°C to 8.85 at 0°C [Blanchard, 1984]). An experimental solution should be tested for its pH after all the components (e.g. EDTA, DTT, Mg^{2+}) have been added since the pH may change after their addition.

- Unless other instructions are given, assume that the pH of a buffer is adjusted down with HCl and up with either NaOH or KOH.

- The basicity of tetramethylammonium hydroxide is equivalent to NaOH or KOH. Tetramethylammonium hydroxide should be used in adjusting the pH of a buffer for a reaction which requires the complete absence of mono-, di-, or trivalent metal ions (Calbiochem "Buffers" booklet).

- By convention, the molarity of a buffer corresponds to the acid component. Thus, 1 M Tris-acetate, pH 4.8 means that 1 M acetic acid is titrated with about 0.5 M Tris base to give the final pH. However, in common usage, just the opposite is often intended: a solution of Tris base may be titrated with acetic acid. Or, even more confusingly, Tris-HCl (the "acid" form) may be titrated with acetic acid, resulting in a three-component system of Tris, chloride, and acetate. Be aware of this potential for ambiguity when trying to duplicate recipes from the literature.

- If both protonated and unprotonated forms of a buffer are readily available, solutions of the two forms at the same concentration can be mixed until the desired pH is obtained. It is preferable to use established tables or calculations for mixing components, although verification with a pH meter is always advisable as in the following example:

Preparation of a phosphate buffer between pH 5.8 and 7.8 (Calbiochem "Buffers" booklet):

Stock Solution A (0.2 M NaH_2PO_4):
   Dissolve 27.6 g NaH_2PO_4 to make 1 liter in deionized water.

Stock Solution B (0.2 M Na_2HPO_4):
   Dissolve 28.4 g Na_2HPO_4 to make 1 liter in deionized water.

pH	% A	% B	pH	% A	% B
5.8	92.0	8.0	6.9	45.0	55.0
5.9	90.0	10.0	**7.0**	**39.0**	**61.0**
6.0	**87.7**	**12.3**	7.1	33.0	67.0
6.1	85.0	15.0	7.2	28.0	72.0
6.2	81.5	19.5	7.3	23.0	77.0
6.3	77.5	22.5	7.4	19.0	81.0
6.4	73.5	26.5	**7.5**	**16.0**	**84.0**
6.5	**68.5**	**31.5**	7.6	13.0	87.0
6.6	62.5	37.5	7.7	10.5	89.5
6.7	56.5	43.5	7.8	8.5	91.5
6.8	51.0	49.0			

Note: pH values are approximate and will vary according to exact temperature and the accuracy at which components are mixed and measured. Therefore, they should be verified with a pH meter. However, pH electrodes themselves are subject to significant variability. Thus, for maximum reproducibility, it is best to adjust pH by adding known amounts of each component, rather than titrating each time with a pH meter.

Table 1.1
pK$_a$ Values of Common Biological Buffers

Trivial Name	Buffer Name	pK$_a$*
Phosphate (pK$_{a1}$)		2.15
Citrate (pK$_{a1}$)		3.06
Formate		3.75
Succinate (pK$_{a1}$)		4.21
Citrate (pK$_{a2}$)		4.76
Acetate		4.76
Pyridine		5.23
Citrate (pK$_{a3}$)		5.40
Succinate (pK$_{a2}$)		5.64
MES	2-(N-Morpholino)ethanesulfonic acid	6.15
Cacodylate	Dimethylarsinic acid	6.27
Carbonate (pK$_{a1}$)		6.35
BIS-Tris	[Bis-(2-hydroxyethyl)imino]tris(hydroxymethyl)methane	6.46
ADA	N-2-Acetamidoiminodiacetic acid	6.59
PIPES	Piperazine-N,N'-bis(2-ethanesulfonic acid)	6.76
BIS-Tris propane	1,3-Bis[tris(hydroxymethyl)methylamino]propane	6.80
ACES	N-2-Acetamido-2-aminoethanesulfonic acid	6.90
Imidazole		6.95
MOPS	3-(N-Morpholino)propanesulfonic acid	7.20
Phosphate (pK$_{a2}$)		7.20
TES	2-[Tris(hydroxymethyl)methylamino]ethanesulfonic acid	7.50
HEPES	N-2-Hydroxyethylpiperazine-N'-2-ethanesulfonic acid	7.55
HEPPS (EPPS)	N-2-Hydroxyethylpiperazine-N'-3-propane-sulfonic acid	8.00
Tris	Tris(hydroxymethyl)aminomethane	·8.06
Tricine	N-[Tris(hydroxymethyl)methyl]glycine	8.15
Glycylglycine		8.25
Bicine	N,N-Bis(2-hydroxyethyl)glycine	8.35
TAPS	3-{[Tris(hydroxymethyl)methyl]amino}propane-sulfonic acid	8.40
Borate		9.23
Ammonia		9.25
CHES	Cyclohexylaminoethanesulfonic acid	9.55
Glycine		9.78
Carbonate (pK$_{a2}$)		10.33
CAPS	3-(cyclohexylamino)propanesulfonic acid	10.40
Phosphate (pK$_{a3}$)		12.43

* Values from Calbiochem "Buffers" booklet or Blanchard (1984).

C. Concentration Effects of Buffer on pH

• It is useful to prepare buffers as 10x or 100x stocks. This permits smaller storage volumes and the addition of bactericidal agents such as 0.02% sodium azide which are diluted to insignificant levels before use (Scopes, 1982). Saturating solubilities of some buffers at 0°C (for full chemical names, see Table 1.1):

MES	0.65 M
PIPES	2.3 M
MOPS	3.0 M
TES	2.6 M
HEPES	2.3 M
Tris	2.4 M
Phosphate	2.5 M (as K^+ salt)

• Note that dilution of concentrated stock buffer solution may change the pH. For example, a buffer with 0.1 M NaH_2PO_4 and 0.1 M Na_2HPO_4 is pH 6.7. Tenfold dilution raises the pH to 6.9 while after one hundredfold dilution it is 7.0 (Tipton and Dixon, 1979).

• The pH of Tris decreases by 0.1 unit per tenfold dilution (Calbiochem "Buffers" booklet).

D. Limitations of Certain Buffers

Buffers are often present at the highest concentration of all components in a protein solution and may have significant effects on a protein or enzyme. Buffers composed of inorganic compounds (phosphate, borate, bicarbonate) may interact with enzymes (or their substrates), affecting their activities. Most seriously, some buffers form coordination complexes with di- and trivalent metal ions resulting in proton release, lower pH, chelation of the metal, and formation of insoluble complexes. Buffers with low metal binding constants such as PIPES, TES, HEPES, and CAPS are preferred for studying enzymes with metal requirements (Blanchard, 1984).

- Phosphate:
  1. is a feeble buffer in the pH range 8 - 11;
  2. precipitates or binds many polyvalent cations;
  3. inhibits a large variety of enzymes, including kinases, phosphatases, dehydrogenases, and other enzymes with phosphate esters as substrates (Blanchard, 1984);
  4. exhibits a dependence of pK on buffer dilution (see Section C).

- Citrate binds to some proteins and forms metal complexes (Scopes, 1982).

- Cacodylate is toxic (Scopes, 1982).

- Carbonate has limited solubility and, since it is in equilibrium with CO_2, studies must be carried out in a closed system (Blanchard, 1984).

- ADA absorbs light at wavelengths up to 260 nm and binds metal ions (Good et al., 1966).

- MOPS interferes with the Lowry protein assay, but not with either the Bradford or Bicinchoninic Acid assays (see Chapter 3).

- HEPES:
  1. interferes with the Lowry protein assay, but not with either the Bradford or Bicinchoninic Acid assays (see Chapter 3);
  2. as for all piperazine-based Good buffers (HEPES, EPPS, PIPES; see Good et al., 1966) forms radicals under various conditions and should be avoided in systems where redox processes are being studied (Grady et al., 1988).

- Tris:
  1. is a poor buffer below pH 7.0;
  2. possesses a potentially reactive primary amine;
  3. participates in various enzymatic reactions such as that catalyzed by alkaline phosphatase;
  4. passes through biological membranes (Calbiochem "Buffers" booklet);
  5. is affected by buffer concentration and temperature (see Section C above).

- Borate forms complexes with mono- and oligosaccharides, nucleic acids and pyridine nucleotides, and glycerol (Blanchard, 1984).

E. Preventing Buffer Contamination

• Phosphate-buffered solutions are highly susceptible to microbial contamination. However, 1 M phosphate stock solutions do not usually become contaminated with bacteria (Schleif and Wensink, 1981).

• Filtering the buffer through a sterile ultrafiltration device may be useful for preventing bacterial or fungal growth, especially at pH 6 - 8 (Blanchard, 1984).

• To prevent buffer contamination during storage, 0.02% (3 mM) sodium azide is often used. Sodium azide does not interact significantly with proteins at this concentration. Note: Sodium azide should be used with caution in the presence of heavy metal cations, as it can dry down into highly unstable salt crystals (Rozycki and Bartha, 1981).

• Refrigeration helps to reduce buffer contamination.

F. Water Purity

Water is the primary ingredient in almost every laboratory solution. Most contaminating substances are removed by distillation and deionization, but traces of some compounds sometimes remain and reliable measurements with protein solutions may be affected. A description of various treatment systems for high-level purification of water for laboratory research can be found in Ganzi (1984).

III. Salts, Metal Ions, and Chelators

A. Ionic Strength: 0.1 - 0.2 M KCl or NaCl simulates physiological conditions for many applications (O'Sullivan and Smithers, 1979).

B. Divalent Cations

If a complex is formed between the buffer and a divalent cation such as Ca^{2+} or Mg^{2+}, the capacity for buffering hydrogen ions is reduced. In addition, the availability of the metal ions to participate in an enzymatic reaction may be diminished. Thus, beware of buffers with affinities for metals.

1. Avoid Tris buffers when a metal cofactor is required for protein activity or stability. In 100 mM Tris with 2 mM Mn^{2+}, 29% of the metal is chelated (Morrison, 1979).

2. For purposes of reproducibility, if working with an ATP-binding enzyme, add Mg^{2+} in 1 mM excess over ATP to ensure that essentially all ATP is present as Mg·ATP (Watts, 1973).

C. Chelators

When it is necessary to limit metal effects, specific metal ion chelators should be used. Metal ion chelators also inactivate metalloproteases.

1. To eliminate trace amounts of heavy metals in buffers, 0.1 - 5 mM ethylene diamine tetraacetic acid (EDTA) is commonly used (Scopes, 1982).

2. The most commonly used chelating agents are EDTA and ethylene bis(oxyethylenenitrilo)tetraacetic acid (EGTA). While EDTA displays strong and nonspecific affinity for a variety of metals, the affinity of EGTA for calcium is significantly higher than its affinity for magnesium, permitting the preferential sequestering of calcium in solutions with EGTA (Blanchard, 1984).

3. *o*-Phenanthroline chelates zinc, while *m*-phenanthroline does not (Todhunter, 1979).

IV. Reducing Agents

A. General Considerations

Within the cell, various reducing compounds, notably glutathione, prevent protein oxidation. Once the cell has been disrupted, care must be taken to counteract effects due to increased contact with oxygen and dilution of naturally occurring reducing agents. Many proteins lose activity when oxidized, although this activity may sometimes be restored by reduction of critical thiol groups. The presence of divalent cations may accelerate the formation of disulfide bonds (Scopes, 1982).

B. Specific Recommendations

- 2-Mercaptoethanol, which is easy to use since it may be stored as a solution at 4°C, must be used at a concentration of 5 - 20 mM. Within 24 hours of its introduction into the buffer, 2-mercaptoethanol becomes oxidized, after which it may accelerate protein inactivation (Scopes, 1982).

- Dithiothreitol (DTT or Cleland's reagent) is supplied as a powder and must be stored at -20°C as a stock solution. DTT may be used at 0.5 - 1.0 mM, and oxidation results in the formation of a stable intramolecular disulfide which does not endanger protein sulfhydryls.

- A good strategy is to use 2-mercaptoethanol at a 1:1000 dilution (about 12 mM) during a protein preparation, but 1 - 5 mM DTT for long term storage (Scopes, 1982).

- Note that certain enzymes are sensitive to reduction (Schleif and Wensinek, 1981).

V. Detergents

A. Introduction

Detergents are used most often for the extraction and purification of membrane proteins, which otherwise are usually insoluble in aqueous solution. A number of classes of detergents may be used and some general guidelines for the solubilization and stabilization of membrane proteins are presented in this section.

Detergents are amphiphilic molecules with substantial solubility in water. With the exception of bile salts, the hydrophobic portion of the molecule usually consists of a linear or branched hydrocarbon "tail" whereas the hydrophilic "head" may have one of a number of very different chemical structures. An important property of detergents is the formation of micelles, which are clusters of detergent molecules in which the hydrophilic head portions face outward. Solubilized membrane proteins form mixed micelles with detergent. The hydrophobic (or transmembrane) domain of the protein is shielded from contact with the aqueous buffer by detergent molecules. The critical micelle concentration (CMC) is defined as the lowest detergent concentration at which micelles form. A detergent with a high CMC (e.g., octyl glucoside) will return to the monomeric state upon dilution below this concentration, thus permitting rapid removal of the detergent by dialysis. In addition, the total molecular weight of the protein plus the micelle may be important for dialysis, gel filtration chromatography, and electrophoresis under non-denaturing conditions. Factors affecting the CMC include temperature, pH, ionic strength, presence of multivalent ions or organic solvents, and detergent purity.

Although the presence of high concentrations of detergent often results in protein denaturation, sometimes the subsequent removal of detergent allows the protein to renature. If protein denaturation is a problem, nonionic detergent concentrations below 0.1% are not usually harmful to proteins (Scopes, 1982).

B. Classes of Detergents

Consult Table 1.2 for specific properties of detergents.

• Ionic Detergents

Ionic detergents contain head groups with either positive charges (cationic detergents) or negative charges (anionic detergents). Ionic detergents have the disadvantage of being highly denaturing. However, they permit the separation of proteins into their monomeric forms, facilitating molecular weight determination.

1. Sodium or Lithium Dodecyl Sulfate (SDS, LiDS)

Of these two anionic detergents, lithium dodecyl sulfate (LiDS) has the advantage that it is soluble at 4°C while sodium dodecyl sulfate (SDS) is not. As much as 10 mg or more of LiDS may be necessary for complete solubilization of one mg of membrane protein (Boehringer Mannheim Catalog). Note that the use of these detergents in the presence of potassium buffers or ammonium sulfate may lead to their precipitation at room temperature. The critical micelle concentration is dramatically affected by the salt concentration; for SDS, the CMC drops from 8 mM in the absence of salt to 0.5 mM in 0.5 M NaCl (Helenius et al., 1979).

2. Sodium Cholate and Sodium Deoxycholate

Sodium cholate and sodium deoxycholate (DOC) are anionic detergents that are less denaturing than other ionic detergents (Harlow and Lane, 1988). Two classes of micelles, termed primary (containing up to 9 molecules) and secondary (9 - 60 molecules), occur above the critical micelle concentration. In contrast to other detergents, free cholate or deoxycholate monomers continue to accumulate above the CMC. The pK_a of these detergents is between 8 and 9, and precipitation of the detergent in its acid form is a problem below pH 7.5. Divalent cations cause DOC precipitation.

• Nonionic Detergents

Nonionic detergents have uncharged hydrophilic head groups. As a result, they are less likely to disrupt protein-protein interactions and are particularly useful for isolating functional protein complexes. Nonionic detergents are far less denaturing than ionic detergents; however, protein aggregation may occur in the presence of these detergents.

1. Triton X-100 (Polyoxyethylene [9-10] *p-t*-octyl phenol)

Many proteins retain their activity in 1 - 3% Triton X-100. A tenfold or greater excess of Triton X-100 to membrane lipid (w/w) may be required to solubilize membrane proteins. Triton X-100 has a strong absorbance at 280 nm.

2. Triton X-114 (Polyoxyethylene [7-8] *p-t*-octyl phenol)

Triton X-114 (2%) added to a protein solution has the property of causing a separation between the detergent and the aqueous phases at temperatures above 20°C, its cloud point. Hydrophilic proteins remain in the aqueous phase while integral membrane proteins may be recovered in the detergent phase (Bordier, 1981).

3. Octyl glucoside (1-O-n-octyl-β-D-glucopyranoside)

Octyl glucoside (OG) is better defined chemically than Triton and therefore results may be more reproducible with this detergent. 20 - 45 mM OG is often sufficient to solubilize membrane proteins. OG has a high CMC (see Table 1.2) and is more easily removed from solution than is Triton.

4. Tween 20 (PEG [20] sorbitan monolaurate)

Tween 20 is commonly used to block nonspecific protein interactions in solid phase immunochemistry (ELISA, RIA, immunoblotting). Tween 20 has a very low CMC.

- Zwitterionic Detergents

    Zwitterionic detergents contain head groups with both positive and negative charges. This class of detergents is more efficient than nonionic detergents at overcoming protein-protein interactions while causing less protein denaturation than ionic detergents.

    1. **CHAPS**   (3-[(Cholamidopropyl)dimethyl-ammonio]-1-propanesulfonate)

        CHAPS does not interfere with ion exchange chromatography or isoelectric focusing.   Proteins in solutions containing CHAPS may be frozen safely.

    2. **Zwittergent 3-14**

Table 1.2

Characteristics of Common Detergents

Detergent	Molecular Weight	Critical Micelle Concentration	Concentration for Solubilization	Micelle Size
Sodium Dodecyl Sulfate	288.5[d]	7 - 10 mM, 0.23%[d]		18 kd
Lithium Dodecyl Sulfate	272.4	6 - 8 mM, 0.2%[a]	>10 mg/mg protein[a]	
Sodium Cholate	431[f]	3 - 10 mM, 0.2%[f]		1.8 kd
Sodium Deoxycholate	433[f]	1 - 2 mM, 0.57%[f]		4.2 kd
Triton X-100	about 628[a]	0.3 mM, 0.02%[c]	0.2 - 0.6 mg/mg protein[b]	90 kd
Triton X-114	about 543[d]	0.35 mM, 0.02%	see text (section V.B.)	
Octyl glucoside	292.4[a]	15 - 25 mM,0.5%[a]	20 - 45 mM[a]	8.0 kd
Tween 20	1230[f]	60 µM, 0.006%[f]		76 kd
Tween 80	1310[f]	10 µM, 0.0013%[f]		49 kd
Brij 35	1200[f]	90 µM, 0.01%[e]		
CHAPS	614.9[d]	4 - 8 mM, 0.5%[f]	6 - 10 mM[a]	6.0 kd
Zwittergent 3-14	364[d]	0.3 mM, 0.011%		30 kd

a - Boehringer Mannheim Catalog
b - Helenius and Simons, 1975
c - Hjelmeland and Chrambach, 1984
d - Calbiochem Catalog
e - Helenius et al., 1979
f - Harlow and Lane, 1988

C. Protocol for Membrane Protein Solubilization

• Experimental Steps (from Hjelmeland and Chrambach, 1984):

1. Prepare crude membrane fraction at a protein concentration of about 10 mg/ml in 50 mM buffer, 0.15 M KCl at 4°C.

2. Prepare detergent stock solution (10%, w/v) in the same buffer.

3. Make dilutions of the crude membrane fraction (about 5 mg/ml) in buffer containing the following amounts of detergent: 0.01%, 0.03%, 0.1%, 0.3%, 1%, 3%.

4. Stir gently for 1 hr at 4°C (avoid foaming or sonication).

5. Centrifuge at 100,000 x g for 1 hr at 4°C.

6. Remove supernatant and resuspend pellet in an equal volume of buffer containing the same detergent concentration.

7. Determine the protein concentration of each fraction (see Chapter 3).

8. Determine the enzyme activity in each fraction.

• General Comments

1. The detergent concentration which yields the highest soluble protein content and activity should be used for further studies. If unsatisfactory results are obtained with a variety of different detergents, mixtures of detergents may be tried. Low yields of protein activity may be improved by the addition of glycerol (25 - 50%, v/v), reducing agents (1 mM DTT or 5 mM β-mercaptoethanol), chelating agents (1 mM EDTA), or protease inhibitors (75 μg/ml PMSF, 20 μg/ml leupeptin or pepstatin; see Section VII below).

2. Sometimes, solubilization with phosphate (0.1 - 0.2 M) is successful if KCl solubilization does not work.

3. Protein solubilization very often occurs near the detergent CMC (Hjelmeland and Chrambach, 1984).

4. For separating soluble protein from integral membrane proteins, the method of Bordier (1981) using the nonionic detergent Triton X-114 may be employed. Extraction and separation in Triton X-114 may be used as a first step in the purification of a membrane protein (this step rapidly separates the membrane proteins from lysosomal proteases). Detergent exchange may be performed during ion-exchange chromatography.

5. The interactions of membrane-associated proteins with membranes may also be disrupted by exposing the membrane preparation to conditions of high salt (0.15 - 2.0 M KCl), high or low pH, high doses of chelating agents (10 mM EDTA or EGTA), or denaturing agents such as urea or guanidine HCl (6 - 10 M) (van Renswoude and Kempf, 1984).

6. The definition of a soluble protein is not always clear. The ability of a protein to remain in the supernatant after a one hour centrifugation at 105,000 x g is affected by both the solution density and temperature. Another test that may be applied is gel filtration chromatography (Hjelmeland and Chrambach, 1984).

VI. Protein Environment

A. Surface Effects

Dilute protein solutions often lose activity quickly, possibly via denaturation on surfaces such as glassware, but this effect can be prevented by inclusion of high levels of another protein, commonly bovine serum albumin (BSA). Ideally, to avoid introducing a "contaminating" protein, dilute protein solutions should be rapidly concentrated. However, since enzyme reactions are sometimes assayed with protein concentrations as low as 1 µg/ml which may lead to rapid inactivation, addition of BSA may be necessary. In addition, loss of the purified protein due to nonspecific adhesion onto glass surfaces (1 µg of protein is absorbed on 5 cm² of a glass surface) is significantly diminished when the solution is supplemented with BSA. At least 0.1 mg/ml BSA should be used in assay mixtures, while stored protein may contain as much as 10 mg/ml BSA (Scopes, 1982).

B. Temperature

An enzyme's reaction velocity roughly doubles with a temperature increase of 10°C (for example between 18°C and 28°C) although some "cold-labile" proteins are effectively inactivated at low temperature (i.e., mitochondrial ATPase). Above 30 - 40°C, proteins vary widely in their stability, most becoming inactivated, but some remaining stable even upon boiling (i.e., bacterial alkaline phosphatase).

C. Storage

As a rule, a protein's half-life is extended by storage at low temperatures. Whether the best storage conditions are at 4°C, -20°C, -80°C, or in liquid nitrogen (-200°C) depends on the protein and its intended use.

For short-term storage (one day to one week), the protein can be stored at 4°C in solution if 1) functional activity (enzyme assay, ligand binding, etc.) does not decrease during this period, 2) no degradation is visible by 2-D electrophoresis (Chapter 7), and 3) no significant amount of pelleted material is obtained by centrifuging 10 min at 20,000 x g (which would indicate denaturation).

For long-term storage (more than one week), several options are available depending on the stability of the individual protein. It may be advisable to test several of these methods before committing a large amount of your protein to any one of them:

1. One of the simplest methods is to store the protein at 4°C as a suspended precipitate in ammonium sulfate (Chapter 4, Section II.A.).

2. Alternatively, ammonium sulfate precipitated protein can be pelleted by centrifuging 20 min at 20,000 x g and then frozen in liquid nitrogen and stored at a temperature of -80°C or less.

3. If you want to freeze your protein, it is important to do it in such a way as to minimize possible phase separations which can precipitate or locally concentrate buffers and allow pH drastic pH fluctuations (+/-3 pH units). High protein concentrations (> 2mg/ml) provide some auto-buffering capacity, and freezing in liquid nitrogen reduces the time needed for freezing (Scopes, 1982). The proper method for freezing a protein in liquid nitrogen is to drop it in 50 μl droplets into liquid nitrogen from a Pasteur pipette or an automatic pipettor. Do this slowly, so individual droplets do not fuse. For large volumes, use a peristaltic

pump to automate dropping. When finished, scoop pellets out of the liquid nitrogen and quickly pour them into a storage tube so that the droplets do not thaw. Store at -80°C or in liquid nitrogen. Storing as droplets allows you to "pour" out as much protein as you need for an experiment without subjecting it to multiple freeze-thaw cycles which can be damaging.

4. A last option is to store the protein as a lyophilized powder (see Chapter 4). For many proteins, lyophilization is the safest and most convenient method of long term storage, but for a significant minority it can cause serious problems of inactivation and inability to resolubilize. Again, it is advisable to test this method on a small amount of protein before committing your entire stock.

As a general note, glycerol is often useful in stabilizing proteins. 50% glycerol is often recommended for storage, while working buffers may contain 20 - 30% glycerol without creating problems for handling (Scopes, 1982). Storage in glycerol allows for maintenance of the protein solution at very low temperatures without freezing.

VII. Protease Inhibitors

Disruption of cells for isolation of proteins may also result in the release of proteases from subcellular compartments. These proteases often need to be removed rapidly to ensure that the protein of interest remains intact. Until the protein of interest is purified, protease inhibitors should be used to retard proteolysis. Five commonly used protease inhibitors are listed below, along with indications for their use. Since protease sensitivity varies widely among different proteins, the indicated working concentration may need to be modified. Special care must be taken when diluting the protease inhibitor into the experiment buffer; since many protease inhibitors are poorly soluble in aqueous solution, they must be mixed in thoroughly and rapidly to minimize precipitation. A more complete list of proteases and protease inhibitors can be found in the Boehringer Mannheim catalog or Harris (1989).

A. Common Inhibitors

- PMSF (phenylmethylsulfonyl fluoride):
    1. inhibits serine proteases (e.g., chymotrypsin, trypsin, thrombin) and thiol proteases (e.g., papain);
    2. soluble in isopropanol to 10 mg/ml;
    3. stock solution stable over a year at room temperature;
    4. working concentration: 17 - 174 µg/ml (0.1 - 1.0 mM);
    5. unstable in aqueous solution, add fresh PMSF at every isolation or purification step.

- EDTA (ethylenediamine tetraacetic acid):
    1. inhibits metalloproteases;
    2. soluble in water to 0.5 M at pH 8 - 9;
    3. stock solution stable for over 6 months at 4°C;
    4. working concentration: 0.5 - 1.5 mM (0.2 - 0.5 mg/ml);
    5. add NaOH to adjust pH of stock solution, otherwise EDTA remains insoluble.

- Pepstatin A:
  1. inhibits acid proteases such as pepsin, renin, cathepsin D, and chymosin;
  2. soluble in methanol to 1 mg/ml;
  3. stock solution stable for 1 week at 4°C or 6 months at -20°C;
  4. working concentration: 0.7 µg/ml (1 µM);
  5. insoluble in water.

- Leupeptin:
  1. inhibits serine and thiol proteases such as papain, plasmin, and cathepsin B;
  2. soluble in water to 10 mg/ml;
  3. stock solution stable for 1 week at 4°C or 6 months at -20°C;
  4. working concentration: 0.5 µg/ml (1 µM).

- Aprotinin:
  1. inhibits serine proteases such as plasmin, kallikrein, trypsin, and chymotrypsin;
  2. soluble in water to 10 mg/ml, adjust to pH 7 - 8;
  3. stock solution stable for 1 week at 4°C or 6 months at -20°C;
  4. working concentration: 0.06 - 2.0 µg/ml (0.01-0.3 µM);
  5. avoid repeated freeze-thawing;
  6. inactive at pH >12.8.

B. A Sample Broad Range Protease Inhibitor Cocktail

35 µg/ml PMSF	- serine proteases
0.3 mg/ml EDTA	- metalloproteases
0.7 µg/ml Pepstatin A	- acid proteases
0.5 µg/ml Leupeptin	- broad spectrum

(from Boehringer Mannheim publications)

VIII. References

Blanchard, J.S. 1984. Meth. Enzymol. 104: 404-414. Buffers for Enzymes.

Boehringer Mannheim Catalog: Biochemica Information. 1987. J. Keesey, ed. Boehringer Mannheim Biochemicals, Indianapolis.

Bordier, C. 1981. J. Biol. Chem. 256: 1604-1607. Phase Separation of Integral Membrane Proteins in Triton X-114 Solution.

Calbiochem. Buffers: A Guide for the Preparation and Use of Buffers in Biological Systems. D.E. Gueffroy, ed. 1975.

Calbiochem Biochemical/Immunochemical Catalog. 1989.

Ellis, K.J. and J.F. Morrison. 1982. Meth. Enzymol. 87: 405-426. Buffers of Constant Ionic Strength for Studying pH-Dependent Processes..

Fink, A.L. and M.A. Geeves. 1979. Meth. Enzymol. 63: 336-370. Cryoenzymology: The Study of Enzyme Catalysis at Subzero Temperatures.

Ganzi, G.C. 1984. Meth. Enzymol. 104: 391-403. Preparation of High-Purity Laboratory Water.

Good, N.E., G.D. Winget, W. Winter, T.N. Connolly, S. Izawa and R.M.M. Singh. 1966. Biochem. 5: 467-477. Hydrogen Ion Buffers for Biological Research.

Grady, J.K., N.D. Chasteen and D.C. Harris. 1988. Anal. Biochem. 173: 111-115. Radicals from "Good's" Buffers.

Harlow, E. and D. Lane. 1988. Antibodies: A Laboratory Manual. 726 pages. Cold Spring Harbor Laboratory, Cold Spring Harbor, New York.

Harris, E.L.V. 1989. Initial Planning. in Protein Purification Methods: A Practical Approach. Harris, E.L.V. & S. Angal, eds. pp.1-66. IRL Press, Oxford.

Helenius, A. and K. Simons. 1975. Biochim. Biophys. Acta 415: 29-79. Solubilization of Membranes by Detergents.

Helenius, A., D.R. McCaslin, E. Fries and C. Tanford. 1979. Meth. Enzymol. 56: 734-749. Properties of Detergents.

Hjelmeland, L.M. and A. Chrambach. 1984. Meth. Enzymol. 104: 305-318. Solubilization of Functional Membrane Proteins.

Morrison, J.F. 1979. Meth. Enzymol. 63: 257-294. Approaches to Kinetic Studies on Metal-Activated Enzymes.

O'Sullivan, W.J. and G.W. Smithers. 1979. Meth. Enzymol. 63: 294-336. Stability Constants for Biologically Important Metal-Ligand Complexes.

van Renswoude, J. and C. Kempf. 1984. Meth. Enzymol. 104: 329-339. Purification of Integral Membrane Proteins.

Rozycki, M. and R. Bartha. 1981. Appl. Environ. Microbiol. 41: 833-836. Problems Associated with the Use of Azide as an Inhibitor of Microbial Activity in Soil.

Schleif, R.F. and P.C. Wensink. 1981. pp. 77-78, 177-178. Practical Methods in Molecular Biology. 220 pages. New York, Springer-Verlag.

Scopes, R.K. 1982. pp. 185-193. Protein Purification: Principles and Practice. 282 pages. Springer-Verlag, New York.

Stryer, L. 1988. Biochemistry. 1089 pages. Third Edition, W.H. Freeman and Company, New York.

Tipton, K.F. and H.B.F. Dixon. 1979. Meth. Enzymol. 63: 183-234 (1979). Effects of pH on Enzymes.

Todhunter, J.A. 1979. Meth. Enzymol. 63: 383-411. Reversible Enzyme Inhibition.

Watts, D.C. 1973. Creatine Kinase. in The Enzymes P.D. Boyer, ed. Vol. 8, pp.383-455. Academic Press, New York.

Chapter 2

Protein Extraction and Solubilization

I. Introduction

II. Protein Extraction
    A. Homogenization
        1. General Considerations
        2. Specific Steps

    B. Sonication
        1. General Considerations
        2. Specific Steps

    C. French Pressure Cell
        1. General Considerations
        2. Specific Steps

    D. Grinding With Alumina or Sand
        1. General Considerations
        2. Specific Steps

    E. Glass Bead Vortexing
        1. General Considerations
        2. Specific Steps

    F. Enzymatic Treatment
        1. General Considerations
        2. Specific Steps

    G. Other Lysis Methods
        1. Detergent Lysis
        2. Organic Solvent Lysis
        3. Osmotic Shock Lysis
        4. Freeze-Thaw Lysis

2

III. Solubilization of Protein from Inclusion Bodies
    A. Solubilization of Inclusion Body Protein
        1. General Considerations
        2. Specific Steps
    B. Prevention of Inclusion Body Formation

IV. Suppliers

V. References

I. Introduction

To purify or characterize an intracellular protein, an efficient method of cell disruption must be developed which releases the protein in soluble form from its intracellular compartment into a solution of well-defined composition. Because this step is the starting point for all subsequent procedures, the disruption method must not be harmful to the protein of interest. Standard procedures for lysis of certain types of tissues or cells are available, but in some cases, it is necessary to explore alternative lysis methods in order to improve protein yields or to maximize recovery of active protein.

Sometimes, cell disruption alone does not release active protein into solution. This is the case for **inclusion bodies**, accumulations of insoluble protein that often form in bacteria which have been induced to express a single protein at very high levels in the cell. In such cases, measures must be taken to convert the protein into soluble, active form at some point prior to, or during, purification of the protein.

Section II discusses the principal techniques of cell lysis. For more detailed explanations, see Methods of Enzymology, vol. 182 (1990). Approaches to solubilizing proteins from inclusion bodies are discussed in Section III.

2

II. Protein Extraction

The first steps of a typical protein isolation procedure usually consist of washing the tissue and applying the lysis method. Then, a centrifugation step separates the soluble proteins from the membrane fraction and insoluble debris. Finally, the protein sample may be analyzed, further purified, or stored.

Tissue is washed with a buffered saline solution to remove traces of blood or other extracellular material, whereas microbial cells may be centrifuged, resuspended in buffer, and recentrifuged to eliminate undesired traces of the growth medium. The extract obtained after lysis, termed the **homogenate**, is usually centrifuged under conditions ranging from 10 minutes at 15,000 x g to one hour at 100,000 x g. The subsequent supernatant is called the **crude extract** and the pellet contains the membrane fraction. If the crude extract contains floating particles, the extract may be filtered through cheesecloth or glass wool. Storage of protein samples is described in Chapter 1 (Section VII.C.).

A list of lysis methods for various tissues or cells is presented in Table 2.1.

Table 2.1

Cell Disruption Methods for Various Tissues

Chapter Section	Cell Lysis Method		Kind of Tissue
II.A	Blade Homogenization		most animal, plant tissues
II.A	Hand Homogenization	(Dounce Homogenization)	soft animal tissues
II.B	Sonication		cell suspensions
II.C	French Pressure Cell		bacteria, yeast, plant cells
II.D	Grinding		bacteria, plant tissues
II.E	Glass Bead Vortexing		cell suspensions
II.F	Enzyme Digestion		bacteria, yeast
II.G.1	Detergent Lysis		tissue culture cells
II.G.2	Organic Solvent Lysis		bacteria, yeast
II.G.3	Osmotic Shock		erythrocytes, bacteria
II.G.4	Freeze-Thaw Lysis		

2

A. Homogenization

1. General Considerations

One of the most common ways of disrupting soft tissues is by homogenization. Homogenization is accomplished either by chopping the tissue in a blender (Lu and Levin, 1978; Necessary et al., 1985) or by forcing the tissue through a narrow opening between a Teflon pestle and a glass container (Fleischer et al., 1979). These methods are rapid and pose relatively little danger to proteins apart from the release of proteases from other cellular compartments.

Time for lysis (from start until centrifugation): 5 - 10 min.

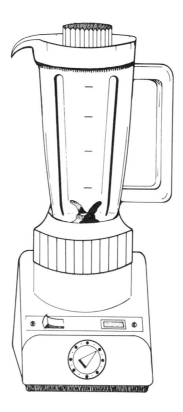

Figure 2.1. Homogenizer.

2. Specific Steps

- **Protocol**

    1. Chop washed tissue into small pieces (for example, 1 cm cubes) with a knife.

    2. Add homogenization buffer, usually 3 - 5 volumes buffer per volume of tissue, and transfer to glass mortar or blender.

    3. Preparation of homogenate:

        a. using a power-driven Potter-Elvehjem glass-Teflon homogenizer at 500 - 1500 rpm, pass through the sample 3 - 6 times, allowing 5 - 10 seconds per stroke.

        b. using a Dounce hand homogenizer, pass through the sample 10 - 20 times.

        c. using a blender, mix three times for 20 seconds at high speed, pausing for several seconds between each pulse.

- **Comments**

    1. Mammalian tissues such as liver, heart, brain, and smooth muscle are commonly disrupted with the Potter-Elvehjem homogenizer. The Dounce hand homogenizer has been used for cultured cells and brain tissue.

    2. The clearance between the pestle and the glass container may be varied to provide more or less shearing of tissue. Clearances may range from 0.35 to 0.70 mm.

    3. The glass container can be immersed in ice water to maintain a low temperature during homogenization. The blender should be prechilled to 4°C, and blending can take place in a cold room.

    4. The extent of cell breakage can be monitored by observing the cells in a phase contrast microscope or by quantifying the amount of protein liberated per gram of tissue wet weight.

    5. If this method does not yield satisfactory lysis, a somewhat harsher technique, such as French pressing or vortexing with glass beads, should be tried.

2

B. Sonication

1. General Considerations

A sonicator disrupts tissue by creating vibrations which cause mechanical shearing of the cell wall. In order to optimize shearing, it is necessary to achieve maximal agitation by tuning the sonicator. Maximal agitation must be tempered by the need to keep the vibrations below the level where foaming of the solution occurs, since this will aerate the solution and cause protein denaturation.

The sonicator probe must be maintained below the solution surface at all times while it is turned on. To tune the sonicator prior to use, place the sonicator probe in an expendable portion of the sample (or a solution with similar viscosity) in the same size vessel as will be used for sonicating the sample. Most commonly, heavy plastic test tubes are used to provide the greatest probe-to-solution surface area. Then, the power is turned on and adjusted to a level slightly below that at which foaming occurs. At this point, the sonicator is considered tuned and only minor adjustments will be necessary during sonication of the sample.

Time for lysis: 5 - 10 min.

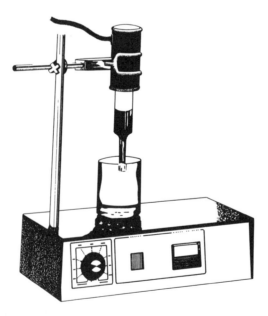

Figure 2.2. Sonicator.

2. Specific Steps

- **Protocol**

    1. Wash the tissue and mince into small pieces with a knife or blender, then suspend in at least 2 volumes of buffer.

    2. Sonicate for about 2 minutes. If the solution starts to foam, decrease the power setting until foaming ceases. Sonication is often carried out for several cycles (e.g. 4 x 30 sec) to permit the sample to cool on ice between treatments.

- **Comments**

    1. Check for cell disruption using phase contrast microscopy after different periods of disruption by sonication.

    2. Up to 1 g of cells or tissue can be lysed at a time with a sonicator.

    3. This method has been used for preparations from many sources, including *Escherichia coli*, *Bacillus subtilis*, *Klebsiella pneumoniae*, and homogenized brain tissue.

    4. Examples for the use of sonication in membrane disruption may be found in Hochstadt (1978), Pederson and Hullihen (1979), and Enquist and Sternberg (1979).

2

C. French Pressure Cell

1. General Considerations

The French Pressure Cell (French Press) achieves cell lysis by subjecting the sample to high pressure followed by a sudden release to atmospheric pressure. The rapid change in pressure causes cells to burst. Most often, the method is used for lysing bacteria and yeast cells. This procedure works very well for moderate volumes (10 - 30 ml), but becomes technically difficult with smaller volumes and too time-consuming with larger volumes. The French pressure cell should be thoroughly cleaned before and after use to prevent contamination due to microbes, and the O-ring controlling the rate of sample release must be replaced regularly due to wear caused by the high pressure.

Time for lysis: 10 - 30 min.

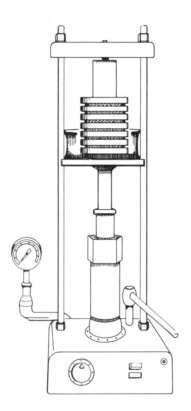

Figure 2.3. French pressure cell.

2. Specific Steps

- **Protocol**

    1. Resuspend the washed cells in lysis buffer. Ratios of cell wet weight to buffer volume range from 1:1 to 1:4 g/ml.

    2. Add sample to the French Pressure cell and bring to desired pressure (commonly 8000 to 20,000 pounds per square inch, or 550 - 1400 kg/cm^2).

    3. While maintaining the pressure, adjust the outlet flow rate to 2 - 3 ml per minute (about one drop every second).

    4. For some samples, a second or even third pass through the French Press is useful for greater lysis.

- **Comments**

    This method has been used for preparations from many sources, including *Escherichia coli*, *Saccharomyces cerevisiae*, *Pseudomonas fluorescens*, *Paracoccus denitrificans*, and *Azotobacter vinelandii* (for examples, see Adair and Jones, 1978, and Schramm and Leung, 1978).

2

D. Grinding With Alumina or Sand

1. General Considerations

As with the French Press, grinding of cells with abrasive materials is an effective means of lysis for unicellular organisms. The materials are inexpensive -- a mortar, pestle, and either sand or alumina -- and the procedure lends itself well to wet cell weights up to 30 g. Cell lysis is achieved by the abrasive action of grinding the thick paste of sample by hand with alumina or sand (Fahnestock, 1979; Sebald et al., 1979).

Time for lysis: 5 - 15 min.

2. Specific Steps

• **Protocol**

1. To a chilled mortar, add 20 g washed cell paste.

2. Prepare 40 g of alumina or quartz sand to add when grinding.

3. Using the pestle, grind cell paste while gradually adding alumina or sand.

4. Grinding should continue for several minutes after the last addition of sand. It is a good sign if grinding results in snapping noises.

5. After grinding, take up cells in 60 ml of buffer and centrifuge to remove cell debris and sand or alumina.

• **Comments**

This method has been used for preparations from many sources, including *Escherichia coli*, *Bacillus subtilis*, *Neurospora crassa*, *Saccharomyces cerevisiae*, as well as plant tissues.

E. Glass Bead Vortexing

1. General Considerations

Vortexing with glass beads is in some ways an extension of the grinding method of cell lysis. The abrasive action of the vortexed beads shears cell walls, liberating the cytoplasmic contents. This method is used primarily on unicellular organisms, particularly yeasts (Schatz, 1979). The method described is for small samples (up to 3 g) and may be carried out in a test tube. For larger samples, specialized apparatus have been developed and they are listed below.

Time for lysis: 10 - 20 min.

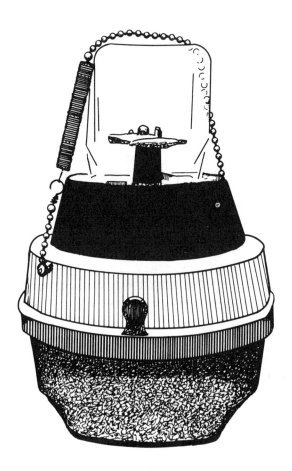

Figure 2.4. Glass-bead homogenizer.

2

2. Specific Steps

- **Protocol**

    1. Resuspend 0.1 - 3 g washed cells in an equal volume of lysis buffer and place in a sturdy tube (preferably not glass since it may shatter due to the action of the glass beads). An Eppendorf tube works well for small samples.

    2. Add 1 - 3 g of chilled glass beads per gram of cell wet weight.

    3. Vortex 3 - 5 times for one minute, each time keeping the cells on ice for one minute between vortexings. Use the highest setting of the vortex mixer.

- **Comments**

    1. To prevent leakage during vortexing, tubes should be closed with a screw cap containing a rubber gasket or should be sealed with Parafilm.

    2. This method is used most frequently with the yeast *Saccharomyces*.

    3. Apparatus which permit glass bead vortexing with larger quantities of cells include the Braun MSK Glass Bead Mill, the Biospec Products Bead-Beater, and the Manton-Gaulin homogenizer.

    4. Glass beads (500 μm diameter) are prepared by washing in concentrated HCl, followed by extensive rinsing (check that the pH is neutral) and drying. Dried glass beads may be chilled prior to use.

F. Enzymatic Treatments

1. General Considerations

Disruption of cells by enzymatic means is principally used with microorganisms, since a relatively uniform treatment is obtained when cells are in suspension.   A protocol for enzymatic disruption of *Escherichia coli* follows.

Time for lysis: 15 - 30 min.

2. Specific Steps

- **Protocol for *E. coli* Cells**

    1. Suspend washed *E. coli* cells in 3 ml TE buffer per gram of cells and bring to 20 - 37°C.

        TE Buffer: 50 mM Tris-HCl (8.0), 10 mM EDTA

    2. Add 1 mg lysozyme per 5 ml of cell suspension (may be added from a freshly made 10 mg/ml stock solution in TE Buffer or directly as lyophilized powder into cell suspension).

    3. Incubate for 10 - 20 min at 20 - 37°C, shaking gently.

- **Comments**

    1. More rapid cell lysis may be obtained by raising the lysozyme concentration (to as much as 10 mg/ml). With higher lysozyme concentrations, satisfactory lysis may be obtained in as little as 5 min even at temperatures as low as 4°C.

    2. A protocol for enzymatic lysis of yeast cells can be found in Harlow and Lane (1988, pp. 455-456) and Jazwinski (1990).

2

G. Other Lysis Methods

1. Detergent Lysis

- **General Considerations**

    Cell lysis with detergents is commonly used with cultured animal cells (Kreibich and Sabatini, 1974; Fong et al., 1982). If low detergent concentrations are sufficient to cause cell lysis, this method may be more gentle to the protein of interest than other lysis methods.

    Time for lysis: 20 - 90 min.

- **Protocol**

    1. Wash cells several times with a buffered saline solution at 4°C (for example, TBS: 10 mM Tris-HCl (pH 7.5), 150 mM NaCl).

    2. After the last centrifugation, resuspend the cells (10^7 - 10^8 cells/ml or 3 - 4 mg total protein/ml) in buffer containing 0.1 - 0.3% Triton X-100.

    3. Vortex or invert tube several times.

    4. Incubate on ice for 10 - 60 min.

- **Comments**

    1. If the protein of interest is unstable, the incubation on ice may be eliminated. This may result in less efficient cell lysis.

    2. To stabilize proteins following extraction, 0.2 volumes of 50% glycerol may be added to the extract.

    3. Detergents other than Triton X-100 may also be used. Consult Chapter 1 for more information about detergents.

2. Organic Solvent Lysis

Lysis of cells using organic solvents is most commonly done with bacteria, though this method has been mainly limited to lysing cells onto filters to be incubated with antibody or nucleic acid probes (Ehrlich et al., 1979; Young and Davis, 1983).

3. Osmotic Shock Lysis

Cells are susceptible to lysis by osmotic shock when suspended in a hypotonic solution (i.e., of a lower ionic strength than the cell cytoplasm) if they are not protected by a cell wall. This method is commonly used for red blood cells.

4. Freeze-Thaw Lysis

It is possible to disrupt cells by subjecting them to several cycles of freezing and thawing. However, this method should only be tried when working with a particularly stable protein which will resist denaturation and is protected from proteolysis.

2

III. Solubilization of Protein from Inclusion Bodies

Bacteria are a particularly convenient source of protein for purification purposes. They can be grown in very large quantities under well-defined conditions, and they are relatively easy to break open for extraction. Most importantly, molecular cloning techniques allow high levels of expression in bacteria of almost any protein from any organism, bacterial or otherwise.

Unfortunately, bacterially expressed proteins are often difficult to purify because of their tendency to precipitate within the cell. The precipitated protein forms inclusion bodies: dense, granular structures which are distributed throughout the cytoplasm (Marston, 1986; Krueger et al., 1989; Hockney, 1995). The formation of inclusion bodies is especially common for non-bacterial proteins, but even native bacterial proteins show a tendency to aggregate when they are expressed at very high cellular levels. Inclusion bodies may also form from bacterial proteins that have been altered by mutating single amino acid residues (Krueger et al., 1990).

Although no single method can be applied to every protein, a number of strategies are available to either solubilize inclusion-body protein directly or to modify the conditions of cell growth and protein expression to minimize precipitation. Some of these strategies are discussed in the following sections.

A. Solubilization of Inclusion Body Protein

1. General Considerations

The exact cause of inclusion body formation is not known. A number of studies have ruled out the possibility that the expressed protein concentration exceeds the native solubility limit of the protein (Mitraki and King, 1989). Sulfhydryl mispairing also appears not to be the primary factor (Kane and Hartley, 1988), even though many proteins with cysteinyl residues do exhibit incorrect (mismatched) disulfide bond pairing (Pigiet and Schuster, 1986). Instead, the prevailing view is that inclusion body formation is linked to the protein folding pathway (Haase-Pettingell and King, 1988; Mitraki and King, 1989; Krueger et al., 1990).

As the nascent polypeptide chain of a protein is synthesized in the cell, it can adopt a number of **intermediate folding conformations** until it finds its correct, **native** three-dimensional structure. Some of the folding intermediates may expose patches of residues which are not normally presented on the surface of the native protein: for example, hydrophobic residues that should be at the core of the protein. When this happens at high concentrations in the cytoplasm, protein intermediates may associate with each other faster than they can refold into their native conformations. This leads to their precipitation into insoluble inclusion bodies.

Although inclusion-body protein may contain a mixture of non-native conformations, the polypeptide chains themselves are usually complete and intact. In fact, purification of protein from inclusion bodies can offer certain advantages, since the bodies are often readily separated from other cell components (Marston, 1986) and can thus be isolated in a relatively pure state.

The rationale in renaturing protein from inclusion bodies is to recognize that the protein folding process must be completed. Since the polypeptide chains in inclusion body protein are trapped in a number of partially or incompletely folded conformations that

2

are stabilized by associations with other polypeptide chains, the first step is to dissociate and solubilize the protein. The protein contained within inclusion bodies is generally insoluble in salts and nonionic detergents, so ionic detergents (1% SDS) or protein denaturants such as urea (8 M) and guanidine hydrochloride (6 M) are used to achieve complete denaturation. Then, the protein is **renatured** by removing the denaturing agent under conditions which favor complete folding of the protein over the formation of intermolecular protein:protein interactions. Some of these conditions include lower temperature and protein concentration than occur in the cytoplasm of growing cells, addition of sulfhydryl reagents to increase disulfide exchange reactions during the folding process, and addition of cofactors and ligands to favor the correctly folded native structure.

2. Specific Steps

- **Protocol**

Diagnosis

Normally, overexpressed proteins can be detected by antibody binding or by observing the presence of protein at the appropriate molecular weight when extracts of expressing cells are compared to nonexpressing control cells. A diagnosis of inclusion body accumulation is usually made when the protein does not appear to be present in soluble portions of cell extracts, even though it may detected at high levels in total cell extracts using denaturing gel electrophoresis (Chapter 5) or immunoblotting (Chapter 8) (van Kimmenade et al., 1988). Inclusion body formation within bacterial cells can also be detected using electron microscopy (Figure 2.5).

Cell Lysis

1. Choose a working buffer for cell lysis. A good general-purpose buffer is 50 mM phosphate, 50 mM NaCl, 1 mM EDTA (pH 7.0).

2. Suspend cells in an appropriate volume of working buffer for the chosen extraction method (Section II). Suitable extraction methods for bacterial cells include sonication, glass bead milling, and lysozyme treatment.

3. Lyse cells as described in Section II. It is generally advisable to carry this step out in the presence of protease inhibitors (Chapter 1).

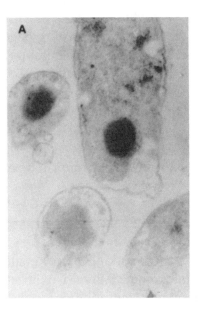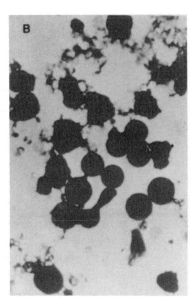

Figure 2.5. Electron micrographs of inclusion bodies. (A) Thin section of *E. coli* showing inclusion bodies of mutant CheB protein. Cells were embedded and negatively stained. (B) Electron micrograph of negatively stained *E. coli* cell extract containing inclusion bodies of CheB. Reprinted with permission from Krueger et al. (1990).

2

Isolation of Inclusion Bodies

4. Centrifuge the cell lysate 5 min at 12,000 x g and 4°C.

5. Resuspend the pellet in 9 volumes of working buffer containing 0.5% of the nonionic detergent Triton X-100. This removes cell membranes and other debris.

6. Incubate at room temperature for 5 min.

7. Recentrifuge.

Solubilization of Inclusion Bodies

8. Suspend pellet in working buffer containing 8 M urea, 2 mM reduced glutathione and 0.2 mM oxidized glutathione. Use a volume of 9 ml/g pellet, but in any case do not exceed a protein concentration of 2.5 mg/ml (Chapter 3). The formation of mixed glutathione-protein disulfides accelerates the correct pairing of disulfides to allow formation of the native structure (Light, 1985).

9. Allow solution to remain at room temperature for 60 min.

Renaturation

10. Slowly add 9 ml of working buffer containing 2 mM reduced glutathione and 0.2 mM oxidized glutathione for each ml of urea-protein solution.

11. Incubate at room temperature for 2 - 4 hr.

• **Comments**

1. Do not forget to assay the lysis supernatant for soluble protein! In many cases, a significant amount of soluble protein persists even when inclusion body formation is extensive.

2. Check for cell disruption using phase contrast microscopy after lysis. Intact cells will sediment with inclusion bodies, diminishing the purity of the inclusion-body protein, so for maximum purity, lysis must be carried out to completion (Krueger et al., 1990).

3. 6 M guanidine hydrochloride or 1% SDS (w/v) can be used instead of 8 M urea for solubilization of inclusion bodies. However, SDS may be difficult to remove in subsequent steps.

4. Instead of using glutathione, disulfide interchange in the refolding protein can be effected by solubilizing the protein at pH 10.7 and then adjusting the pH to 8 by titrating with HCl *after* step 10 (Marston et al., 1984). Disulfide interchange considerations can be ignored if the protein of interest does not contain cysteine residues.

5. Slower refolding can be effected by dialyzing the fully denatured protein against decreasing concentrations of denaturant.

6. Dilution of urea-denatured protein into protein precipitating agents has been found to facilitate renaturation. A sparse matrix method for finding the best renaturation conditions has been described and is based on combinations of buffers and precipitating agents that are also useful in protein crystallization (Hofmann et al., 1995; see also Chapter 12).

2

7. Renaturation may be facilitated by the addition of cofactors, ligands, substrates, or prosthetic groups which may stabilize folding intermediates that are closely related to the native structure of the protein.

8. Examples of renaturation protocols may be found in Marston et al. (1984), van Kimmenade et al. (1988), Krueger et al. (1990), and Hofmann et al., 1995. In some cases, renaturation can be facilitated by chemical modification of residue side chains (Light, 1985).

B. Prevention of Inclusion Body Formation

Because inclusion bodies result from accumulation of incompletely folded polypeptides, an alternative strategy for the purification of bacterially expressed proteins is to modify the conditions of cell growth to either slow the rate of association of partially folded intermediates or to speed up the rate of folding of the polypeptides into their native structure. Aggregation of bacterially expressed protein into inclusion bodies is rarely 100%. More often, the protein shows some degree of partitioning between inclusion bodies and soluble, native protein. Even if complete solubilization is not obtained, a significant shift of protein from the insoluble to the soluble states may mean the difference in being able to obtain a practical purification scheme.

A number of approaches have been used to maximize the solubility of expressed proteins in bacteria (Hockney, et al., 1995). None has been found to be general enough for all applications, so finding the best approach for a particular protein may require experimenting with a number of them, singly or in combination.

- *Decreased Temperature of Cell Growth*: Aggregation of folding intermediates appears to be driven primarily by the association of exposed hydrophobic surfaces. Since hydrophobic interactions tend to decrease at lower temperatures, decreasing the temperature of the growth medium should lower the rate of association of folding intermediates, allowing them more time to refold into native, soluble tertiary structures. This approach was effective in the expression in *E. coli* of human interferons (Schein and Noteborn, 1988). For cells grown at 30°C, the percentage of soluble protein increased 3- to 16-fold over the levels in cells grown at 37°C. Unfortunately, this strategy is limited by the relatively narrow temperature ranges tolerated for efficient cell growth.

- *Co-expression of Molecular Chaperones*: Molecular chaperones are proteins which expend the energy of ATP hydrolysis to maintain partially folded proteins in the soluble state. Prokaryotic chaperones are homologous to those found in eukaryotic cells (GroEL, GroES, DnaK, HtpG), and they can stabilize folding intermediates of eukaryotic proteins expressed in prokaryotic cells. Co-overexpression of the chaperone DnaK increased the percentage of soluble human growth hormone in *E. coli* by up to 87% (Blum et al., 1992), while co-overexpression with the chaperones GroES and GroEL allowed the purification of milligram quantities of pure recombinant p50^{csk} (Amrein et al., 1995). However, much smaller degrees of solubilization are more generally typical (Hockney, 1995).

- *Osmotic Stress*: The cytokinin biosynthetic enzyme dimethylallyl pyrophosphate: 5'-AMP transferase from *Agrobacterium* was expressed in *E. coli* (Blackwell and Horgan, 1991). Under standard conditions, the enzyme forms inclusion bodies in more than 90% of growing cells, but in the presence of sorbitol and glycyl betaine, it is expressed in almost completely soluble form. Sorbitol facilitates the cellular uptake of glycyl betaine, which in turn stabilizes protein structure by "preferential hydration" (Arakawa and Timasheff, 1985). Note: the authors combined this method with that of decreased temperature of cell growth, described above.

- *Fusion Proteins*: Fusion of the gene for thioredoxin to those of a number of mammalian cytokines and growth factors has been successfully used to obtain high levels of expression and solubility (LaVallie et al., 1993).

2

IV. Suppliers

French Pressure Cell: Thomas Scientific, Baxter Scientific Products

Glass Bead Vortexer: Biospec Products, B. Braun Instruments, Gaulin Corporation

Glutathione (oxidized and reduced): Aldrich, Boehringer-Mannheim, Calbiochem, Fluka, ICN, Serva, Sigma

Guanidine Hydrochloride: Aldrich, Boehringer-Mannheim, Calbiochem, Fluka, ICN, Serva, Sigma

Homogenizer: Fisher, Baxter Scientific Products

Sand: Sigma, Thomas Scientific

Sonicator: Fisher

Triton X-100: Aldrich, Boehringer-Mannheim, Calbiochem, Fluka, ICN, Serva, Sigma

Urea: Aldrich, Boehringer-Mannheim, Calbiochem, Fluka, ICN, Serva, Sigma

V. References

Adair, L.B. and M.E. Jones. 1978. Meth. Enzymol. 51: 51-58. Aspartate Carbamyltransferase (*Pseudomonas fluorescens*).

Amrein, K.E., B. Takacs, M. Steiger, J. Molnos, N.A. Flint and P. Burn. 1995. Proc. Natl. Acad. Sci. USA 92: 1048-1052. Purification and Characterization of Recombinant Human p50[csk] Protein-tyrosine Kinase from an *Escherichia coli* Expression System Overproducing the Bacterial Chaperones GroES and GroEL.

Arakawa, T. and S.N. Timasheff. 1985. Biophys. J. 47: 411-414. The Stabilization of Proteins by Osmolytes.

Blackwell, J.R. and R. Horgan. 1991. FEBS Lett. 295: 10-12. A Novel Strategy for Production of a Highly Expressed Recombinant Protein in an Active Form.

Blum, P., M. Velligan, N. Lin and A. Matin. 1992. Bio/Technology 10: 301-304. DnaK-mediated Alternations in Human Growth Hormone Protein Inclusion Bodies.

Ehrlich, H.A., Cohen, S.N. and McDevitt, H.O. 1979. Meth. Enzymol. 68: 443-453. Immunological Detection and Characterization of Products Translated from Cloned DNA Fragments.

Enquist, L. and N. Sternberg. 1979. Meth. Enzymol. 68: 281-298. In Vitro Packaging of λ Dam Vectors and Their Use in Cloning DNA Fragments.

Fahnestock, S.R. 1979. Meth. Enzymol. 59: 437-443. Reconstitution of Active 50S Ribosomal Subunits from Bacillus lichenformis and Bacillus subtilis.

Fleischer, S., J.O. McIntyre and J.C. Vital. 1979. Meth. Enzymol. 55: 32-39. Large-Scale Preparation of Rat Liver Mitochondria in High Yield.

Fong, S.-L., Tsin, A.T.C., Bridges, C.D.B. and Liou, G.I. 1982. Meth. Enzymol. 81: 133-140. Detergents for Extraction of Visual Pigments: Types, Solubilization, and Stability.

Haase-Pettingell, C. and J. King. 1988. J. Biol. Chem. 263: 4977-4983. Formation of Aggregates from a Thermolabile *In vivo* Folding Intermediate in P22 Tailspike Maturation: A Model for Inclusion Body Formation.

Harlow, E. and D. Lane. 1988. Antibodies: A Laboratory Manual. 726 pages. Cold Spring Harbor Laboratory, Cold Spring Harbor, New York.

Hochstadt, J. 1978. Meth. Enzymol. 51: 558-567. Adenosine Phosphoribosyltransferase from *Escherichia coli*.

2

Hockney, R.C. 1995. Trends Biotech. 13: 456-463. Recent Developments in Heterologous Protein Production in *Escherichia coli*.

Hofmann, A., M. Tai, W. Wong and C.G. Glabe. 1995. Anal. Biochem. 230: 8-15. A Sparse Matrix Screen to Establish Initial Conditions for Protein Renaturation.

Jazwinski, S.M. 1990. Meth. Enzymol. 182: 154-174. Preparation of Extracts from Yeast.

Kane, J.F. and D.L. Hartley. 1988. Trends Biotech. 6: 95-101. Formation of Recombinant Protein Inclusion Bodies in *E. Coli*.

Kreibich, G. and Sabatini, D.D. 1974. Meth. Enzymol. 31: 215-225. Procedure for the Selective Release of Content from Microsomal Vesicles without Membrane Disassembly.

Krueger, J.K., M.H. Kulke, C. Schutt and J. Stock. 1989. BioPharm. Manufacturing 3: 40-45. Protein Inclusion Body Formation and Purification.

Krueger, J.K., A.M. Stock, C.E. Schutt and J.B. Stock. 1990. pp. 136-142 in Protein Folding: Deciphering the Second Half of the Genetic Code. Gierasch, L. M. and J. King, eds. 334 pages. American Association for the Advancement of Science, Washington, D.C.

LaVallie, E.R., E.A. DiBlasio, S. Kovacic, K.L. Grant, P.F. Schendel and J.M. McCoy. 1993. Bio/Technology 11: 187-193. A Thioredoxin Gene Fusion Expression System that Circumvents Inclusion Body Formation in the *E. Coli* Cytoplasm.

Light, A. 1985. BioTechniques 3: 298-306. Protein Solubility, Protein Modifications and Protein Folding.

Lu, A.Y.H and W. Levin. 1978. Meth. Enzymol. 52: 193-200. Purification and Assay of Liver Microsomal Epoxide Hydrase.

Methods in Enzymology. 1990. Volume 182. Guide to Protein Purification. 1990. Academic Press, San Diego.

Marston, A.O. 1986. Biochem. J. 240: 1-12. The Purification of Eukaryotic Polypeptides Synthesized in *Escherichia coli*.

Marston, A.O., P.A. Lowe, M.T. Doel, J.M. Schoemaker, S. White and S. Angal. 1984. Bio/Technology 2: 800-804. Purification of Calf Prochymosin (Prorennin) Synthesized in *Escherichia coli*.

Mitraki, A. and J. King. 1989. Bio/Technology 7: 690-697. Protein Folding Intermediates and Inclusion Body Formation.

Necessary, P.C., B. Roberts, P.A. Humphrey, G.M. Helmkamp, Jr., C.D. Turner, A.B. Rawitch and K.E. Ebner. 1985. Anal. Biochem. 146: 372-373. The Cuisinart Food Processor Efficiently Disaggregates Tissues.

2

Pederson, P.L. and J. Hullihen. 1979. Meth. Enzymol. 55: 736-741. Resolution and Reconstitution of ATP Synthesis and ATP-Dependent Functions of Liver Mitochondria.

Pigiet, V.P. and B.J. Schuster. 1986. Proc. Natl. Acad. Sci. USA 83: 7643-7647. Thioredoxin-catalyzed Refolding of Disulphide-containing Proteins.

Schatz, G. 1979. Meth. Enzymol. 56: 40-50. Biogenesis of Yeast Mitochondria: Synthesis of Cytochrome c Oxidase and Cytochrome c1.

Schein, C.H. and M.H.M. Noteborn. 1988. Bio/Technology 6: 291-294. Formation of Soluble Recombinant Proteins in *Escherichia coli* is Favored by Lower Growth Temperature.

Schramm, V.L. and H.B. Leung. 1978. Meth. Enzymol. 51: 263-271. Adenine Monophosphate Nucleosidase from *Azotobacter vinelandii* and *Escherichia coli*.

Sebald, W., W. Neupert and H. Weiss. 1979. Meth. Enzymol. 55: 144-148. Preparation of *Neurospora crassa* Mitochondria.

Van Kimmenade, A., M.W. Bond, J.H. Schumacher, C. Laquoi and R.A. Kastelein. 1988. Eur. J. Biochem. 173: 109-114. Expression, Renaturation and Purification of Recombinant Human Interleukin 4 from *Escherichia coli*.

Young, R.A. and R. W. Davis. 1983. Proc. Natl. Acad. Sci. USA 80: 1194-1198. Efficient Isolation of Genes by Using Antibody Probes.

Chapter 3

3

Protein Concentration Determination

I. Absorbance at 280 nm (A_{280})
    A. Summary
    B. Equipment
    C. Reagents
    D. Protocol
    E. Comments

II. Bradford Assay
    A. Summary
    B. Equipment
    C. Reagents
    D. Protocol
    E. Comments

III. Lowry Assay
    A. Summary
    B. Equipment
    C. Reagents
    D. Protocol
    E. Comments

IV. Bicinchoninic Acid (BCA) Assay
    A. Summary
    B. Equipment
    C. Reagents
    D. Protocol
    E. Comments

V. Briefly Noted: Dot Filter Binding Assay
    A. Summary
    B. Protocol

VI. Interference Table

VII. Suppliers

VIII. References

I. Absorbance at 280 nm (A_{280})

3

A rapid method of determining whether sample solutions contain protein. Most commonly, absorbance is used for generating a protein elution profile after column chromatography.

A. Summary

• Time required: a few minutes.

• Advantages:
1. Direct
2. Rapid
3. Nondestructive

• Disadvantages:
1. Not strictly quantitative, since this assay is based on the strong absorbance of tyrosine, phenylalanine, and tryptophan residues. Different proteins may therefore have widely varying extinction coefficients; if a protein contains no Phe, Tyr, or Trp, it will be undetected. This assay is adequate for crude protein mixtures.
2. Strong interference by nucleic acids

• Range of sensitivity: 0.2 - 2 mg/ml; can measure as little as 0.1 ml (~0.05 mg) in microcuvettes.

• Theory: Wetlaufer (1962)

B. Equipment

• Spectrophotometer (equipped for UV reading)
• Quartz cuvettes
• Pasteur pipettes and pipette bulbs for solution transfer

C. Reagents: experiment buffer (for blank)

D. Protocol

- Single-beam spectrophotometer

    1. With no cuvette present in instrument, set A_{280} to zero.

    2. With experiment buffer in cuvette, read A_{280}, then reset to zero. (This step determines whether the experiment buffer has a significant absorbance.)

    3. Remove experiment buffer and add sample to cuvette, then record absorbance.

- Dual-beam spectrophotometer

    1. With matched, empty cuvettes in machine, set instrument zero.

    2. Add experiment buffer to sample cuvette, leave reference cuvette empty. Record absorbance (This step determines whether the experiment buffer has a significant absorbance.).

    3. Remove experiment buffer from sample cuvette. Add protein solution and experiment buffer to sample and reference cuvettes, respectively, then record absorbance.

E. Comments

- It is a common laboratory shortcut (although a very imprecise one) to assume that an absorbance of 1.0 in a 1 cm cuvette roughly approximates 1 mg/ml of protein. For comparison, measured A_{280} values of a sampling of proteins at 1 mg/ml follow (adapted from Whitaker and Granum, 1980):

Protein	A_{280}(1 mg/ml)
Bovine Serum Albumin	0.70
Ribonuclease A	0.77
Ovalbumin	0.79
γ-Globulin	1.38
Trypsin	1.60
Chymotrypsin	2.02
α-Amylase	2.42

- If absorbance is off scale, the sample can be diluted with buffer and the assay repeated. Alternatively, a cuvette with a shorter path length may be used.

- If the experiment buffer has a high absorbance relative to water, there is some interfering substance in the buffer (see Interference Table, pp. 76 - 78). Moderate buffer absorbance can be balanced by the zero setting, but spectrophotometers have a limited range of sensitivity at higher absorbances due to interference by stray light. The range varies between instruments, but generally it is advisable to avoid a total (buffer plus protein) absorbance higher than 1.5. *Note: setting the absorbance to zero in the presence of experiment buffer does not eliminate the contribution of this buffer to the total absorbance of the protein solution!* See Appendix 4 for an explanation of how to test a spectrophotometer's absorbance linearity range.

- Glass or plastic cuvettes absorb light in the UV range and should not be used for this assay.

- This technique is most widely used for reading column fractions to obtain an estimate of where the protein peak is eluting, but unless a more specific test is performed, it is dangerous to assume that an A_{280} peak reflects eluted protein. For rapid and sensitive protein assays, try the Bradford or Dot Filter Binding Assays.

- Variations of this assay which permit partial correction for interfering substances and protein composition differences include the A_{280}/A_{260} ratio (Warburg and Christian, 1941) and the A_{280}/A_{235} ratio (Whitaker and Granum, 1980).

- An approximate correction when nucleic acid is present (Schleif and Wensink, 1981):

Protein Concentration (mg/ml) = 1.5 x A_{280} - 0.75 x A_{260}

- A graphic representation of this method is presented in the form of a nomograph (Fig. 3.1). If the A_{280}/A_{260} is roughly 2, it can be assumed that the nucleic acid concentration is negligible.

3

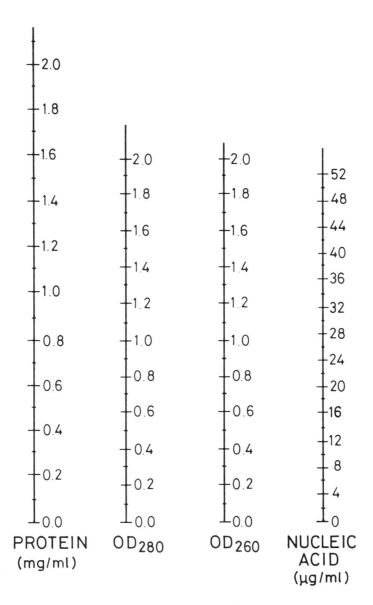

Figure 3.1. Nomograph. Alignment of a straight edge at the two points corresponding to optical density (absorbance) values on the central scales for 260 nm and 280 nm (obtained with a one centimeter light path cuvette) permits the estimated protein and nucleic acid concentrations to be read from the scales on the extreme left and right, respectively. Modified from an earlier version by E. Adams using the data of Warburg and Christian (1941).

II. Bradford Assay

A rapid and reliable dye-based assay for determining protein content in a solution. Although there are relatively few interfering substances, the dye interacts to varying extents with different proteins and thus is not strictly quantitative. This assay is sometimes referred to as the Bio-Rad assay after the company which sells a widely used kit that is based on this method.

A. Summary

- Time Required: 10 min.

- Advantages:
    1. Rapid (two minute development time).
    2. Sensitive, hence little protein must be sacrificed.

- Disadvantages:
    1. Some variability in response between different purified proteins.
    2. Proteins used for this assay are irreversibly denatured.

- Range of sensitivity: 25 - 200 µg/ml protein solution; minimum volume of 0.1 ml permits measurement of as little as 2.5 µg of protein.

- Theory: Bradford, 1976

B. Equipment

- Spectrophotometer
- Plastic cuvettes (polystyrene)
- Pasteur pipettes and pipette bulbs
- Pipettes
- Small disposable test tubes or Eppendorf tubes
- Test tube rack
- Vortex mixer

C. Reagents

- Serva Blue G Dye
- 1 mg/ml bovine serum albumin (BSA)
- 95% ethanol
- 85% phosphoric acid
        or Bio-Rad kit with pre-mixed reagents

D. Protocol

• **Solutions**

    1. Bradford Stock Solution
        100 ml 95% ethanol
        200 ml 88% phosphoric acid
        350 mg Serva Blue G
        Stable indefinitely at room temperature.

    2. Bradford Working Buffer
        425 ml distilled water
        15 ml 95% ethanol
        30 ml 88% phosphoric acid
        30 ml Bradford Stock Solution
        Filter through Whatman No. 1 paper, store at room temperature in brown glass bottle. Usable for several weeks, but may need to be refiltered.

Reagents are also commercially available (Bio-Rad).

• **Assay**

    1. Pipet protein solution (maximum 100 µl) into tube (see example below for standard curve).

    2. Add experiment buffer to make a total volume of 100 µl.

    3. Add 1 ml Bradford Working Buffer and vortex.

    4. Read A_{595} after 2 minutes (Read and Northcote, 1981) but before 1 hour (Bearden, 1978).

3

- **Generating a Standard Curve**

1. A standard curve with samples of known protein concentration prepared in parallel with unknown protein solutions is essential for quantitative assessment of protein concentration.   In addition, samples should be run in duplicate or triplicate. If the same cuvette is used for reading samples, it is a good idea to read those samples with less protein content first to reduce error arising from Bradford dye carryover in the cuvette as a result of incomplete rinsing.

2. Due to the continuing course of color development of the protein-dye complex, the Bradford Working Buffer should be added to standard and unknown protein solutions sequentially. All samples should also be read sequentially following the color reaction development. Each assay of unknown protein solutions must be accompanied by the generation of a new standard curve.

3. The standard curve table on the following page illustrates how to prepare a standard curve. Care must be taken to ensure that the curve remains linear over the range of protein concentrations tested or a large margin of error will be associated with unknown protein concentration readings.

4. After measuring the A_{595}, the curve presented in Fig. 3.2 is generated with the averaged values.

5. The standard curve for the Bradford Assay remains linear only from about 2.5 µg to 15 µg of BSA. If absorbances of unknown protein samples fall outside of this range, the margin of error becomes very high.   Another consequence of the range of linearity is that a line for the standard curve should be drawn with greater emphasis placed on points in the linear range than at the edges. It is also possible to express the amount of BSA measured along the x-axis as a concentration.

Standard Curve

Sample Number	µg protein	Standard solution (1 mg/ml BSA)	Experiment Buffer	Bradford Reagent	A_{595}
1	0 µg	0 µl	100 µl	1 ml	0.000
2	2.5	2.5	97.5	1	0.120
3	2.5	2.5	97.5	1	0.130
4	5	5	95	1	0.250
5	5	5	95	1	0.215
6	7.5	7.5	92.5	1	0.331
7	7.5	7.5	92.5	1	0.364
8	10	10	90	1	0.460
9	10	10	90	1	0.442
10	12.5	12.5	87.5	1	0.531
11	12.5	12.5	87.5	1	0.562
12	15	15	85	1	0.633
13	15	15	85	1	0.617
14	17.5	17.5	82.5	1	0.684
15	17.5	17.5	82.5	1	0.650
16	20	20	80	1	0.721
17	20	20	80	1	0.727

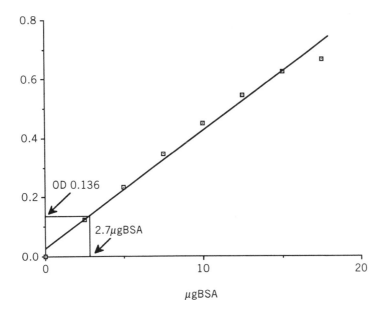

Figure 3.2. Bradford Assay Standard Curve.

3

• Utilizing the Standard Curve

Assuming an unknown protein solution yields the following readings in parallel with the standard curve above:

No.	Sample	Expt Buffer	Bradford Reagent	A_{595}	Av. A_{595}	µg protein	µg/µl (mg/ml)
1	5 µl	95 µl	1 ml	0.059			
2	5	95	1	0.101	}0.082	(unreliable)	
3	5	95	1	0.085			
4	10	90	1	0.135			
5	10	90	1	0.146	}0.136	2.7	0.27
6	10	90	1	0.127			
7	50	50	1	0.700			
8	50	50	1	0.717	}0.718	16.4	0.33
9	50	50	1	0.738			

The calculation of the protein concentration can be accomplished as follows:

1. Determine the average absorbance for a given volume of sample, i.e., for samples 4 - 6, $(0.125 + 0.136 + 0.117)/3 = 0.136$.

2. Using the standard curve, extrapolate the amount of BSA (assume that it equals the amount of protein) in that sample; i.e., for samples 4 - 6, a horizontal line at $OD_{595} = 0.136$ intersects the standard curve at a point which corresponds to 2.7 µg.

3. Calculate the protein concentration by dividing the amount of protein by the sample volume; i.e., for samples 4 - 6, 2.7 µg/10 µl $= 0.27$ µg/µl $(= 0.27$ mg/ml$)$.

4. Ignore the lowest average reading since it is below the linear range of the standard curve.

5. Average $0.27 + 0.33$ to give the overall estimate of 0.30 mg/ml.

3

E. Comments

- Color is fully developed after 5 minutes, but precipitation starts after 10-15 minutes, especially at higher protein concentrations due to the tendency for proteins to precipitate in acidic conditions. If samples are read within 10 minutes of standards, error due to color loss should be <2% (Peterson, 1983).

- Disposable plastic cuvettes are recommended because stain accumulates on cuvette walls and is difficult to remove.

- If glass cuvettes are used, clean them by rinsing thoroughly either with methanol, in concentrated glassware detergent followed by water and acetone, or by soaking in concentrated HCl overnight.

- If sample absorbance is above the useful linear range, dilute with Bradford Working Buffer to a maximum total of 5 ml. The blank should be diluted with an equal volume (Spector, 1978).

- Assay results vary with different purified proteins (Pierce and Suelter, 1977, Van Kley and Hale 1977). Protein concentration measurements may be more accurate when generating the standard curve with the protein of interest or calibrating against another method (i.e., BCA, see Section IV).

- Bovine serum albumin (BSA) is most commonly used as a protein standard, although it has often been pointed out that the response of BSA to the dye is atypical. Ovalbumin is a more representative alternative standard (Read and Northcote, 1981).

- Older reagent solutions produce decreased absorbances (Spector, 1978).

- The Bradford assay can also be used for quantifying proteins immobilized on columns (Asryants et al., 1985).

- This method can be applied to membrane-bound proteins (Fanger, 1987).

III. Lowry Assay

A widely used quantitative assay for determining protein content in a solution (Lowry et al., 1951). Most interfering substances can be removed by precipitating the proteins from solution prior to running the assay.

A. Summary

- Time Required: 40 min.

- Advantages:
    1. Reliable method for protein quantitation
    2. Little variation among different proteins

- Disadvantages:
    1. Many interfering substances
    2. Slow reaction rate
    3. Instability of certain reagents
    4. Proteins irreversibly denatured

- Range of sensitivity: 5 µg/ml - 100 µg/ml

- Theory: Peterson (1983)

B. Equipment

- Spectrophotometer
- Cuvettes
- Pipettes
- Pasteur pipettes and pipette bulbs
- Test tubes (3 - 5 ml capacity)
- Test tube rack
- Vortex mixer

C. Reagents

- $CuSO_4 \cdot 5\ H_2O$
- $Na_3C_6H_5O_7$ ($\cdot 2H_2O$) (Sodium citrate)
- Na_2CO_3
- NaOH
- Folin-Ciocalteu phenol reagent

D. Protocol (from Scopes, 1982)

- Solutions

    1. **Solution A**, 100 ml
        0.5 g $CuSO_4 \cdot 5\,H_2O$

        1 g $Na_3C_6H_5O_7 \cdot 2H_2O$
        Add distilled water to 100 ml
        Solution may be stored indefinitely at room temperature.

    2. **Solution B**, 1 liter
        20 g Na_2CO_3
        4 g NaOH
        Add distilled water to 1 liter
        Solution may be stored indefinitely at room temperature.

    3. **Solution C**, 51 ml
            1 ml Solution A
            50 ml Solution B

    4. **Solution D**, 20 ml
            10 ml Folin-Ciocalteu phenol reagent
            10 ml distilled water

- **Assay**

    1. Bring sample solution to 0.5 ml with distilled water.

    2. Add 2.5 ml Solution C.

    3. Vortex and let stand at room temperature for 5 - 10 minutes.

    4. Add 0.25 ml Solution D and vortex.

    5. After 20 - 30 minutes, read A_{750}.

A sample standard curve is presented in Fig. 3.3.

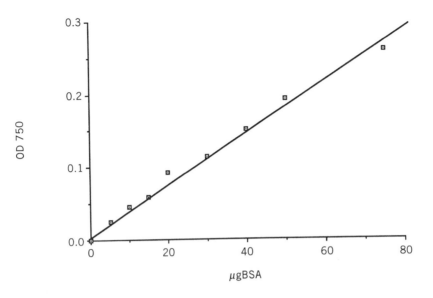

Figure 3.3. Lowry Assay Standard Curve.

- **Additional steps to purify protein sample from interfering substances:**

  Deoxycholate-Trichloroacetic Acid (DOC-TCA) Precipitation also allows concentration determination of proteins in dilute solutions (<1 μg/ml) (from Peterson, 1983).

  Additional Reagents:
  0.15% (w/v) deoxycholate (DOC)
  72% trichloroacetic acid (TCA)

  1. To 1.0 ml protein sample, add 0.1 ml 0.15% DOC.

  2. Vortex and let stand at room temperature for 10 minutes.

  3. Add 0.1 ml 72% TCA, vortex and spin 5 - 30 minutes at 1000 - 3000 x *g*. With fixed angle rotors, cold temperatures or large volumes, longer times are necessary.

  4. Decant immediately and remove residual liquid with a pipette.

  5. Redissolve pellet directly in Solution C.

3

E. Comments (from Peterson, 1977, 1979, 1983)

- Color development reaches a maximum in 20 - 30 minutes, after which there is a gradual loss of signal at about 1% per hour.

- Most interfering substances cause lower color yield, while some detergents cause a slight increase.

- High salt concentrations may cause precipitation.

- Lipids can be removed by chloroform extraction, and centrifugation may remove turbidity due to potassium ions or Triton X-100.

- Interference due to detergents, sucrose, and EDTA may be eliminated by adding SDS to the Lowry reagents (Markwell et al., 1981).

- Extinction coefficients of protein-dye complexes generally vary from one another by a factor of 1.2 or less.

IV. Bicinchoninic Acid (BCA) Assay

A recently developed variation of the Lowry assay. The reaction is simpler to perform and has fewer interfering substances.

A. Summary

- Time Required: 40 min, 2 hrs, or overnight.

- Advantages:
    1. Single reagent
    2. End product is stable
    3. Fewer interfering substances than Lowry assay

- Disadvantages:
    1. Slow reaction time
    2. Proteins irreversibly denatured

- Range of Sensitivity:
    Standard Assay: 10 - 1200 µg/ml
    Microassay: 0.5 - 10 µg/ml

- Theory: Smith et al. (1985)

B. Equipment

- Spectrophotometer
- Water bath at 37°C (optional)
- Cuvettes
- Pipettes
- Pasteur pipettes and pipette bulbs
- Test tubes
- Test tube racks

C. Reagents

- BCA (bicinchoninic acid)
- $Na_2CO_3 \cdot H_2O$ (sodium carbonate)
- $Na_2C_4H_4O_6 \cdot 2H_2O$ (sodium tartrate)
- NaOH (sodium hydroxide)
- $NaHCO_3$
- $CuSO_4 \cdot 5 H_2O$ (copper sulfate)

Available as a kit from Pierce

D. Protocol (from Smith et al., 1985)

- Solutions

    1. Reagent A, 1 liter
        10 g BCA (1%)
        20 g $Na_2CO_3 \cdot H_2O$ (2%)
        1.6 g $Na_2C_4H_4O_6 \cdot 2H_2O$ (0.16%)
        4 g NaOH (0.4%)
        9.5 g $NaHCO_3$ (0.95%)
        Add distilled water to 1 liter
        If needed, add NaOH or solid $NaHCO_3$ to adjust pH to 11.25.

    2. Reagent B, 50 ml
        2 g $CuSO_4 \cdot 5\ H_2O$ (4%)
        Add distilled water to 50 ml

    Reagents A and B are stable for at least 12 months at room
        temperature, and are commercially available (Pierce).

    3. Standard Working Reagent (SWR)
        50 volumes Reagent A
        1 volume Reagent B
        Stable for 1 week (Pierce Bulletin)

- **Assay**

    1. Mix 1 volume of sample with 20 volumes of SWR (i.e. 100 μl
        and 2 ml).

    2. Incubate either (a) at room temperature for 2 hours or (b) at
        37°C for 30 min.

    3. Cool to room temperature in the case of (b).

    4. Read A_{562}.

    A standard curve is presented in Fig. 3.4.

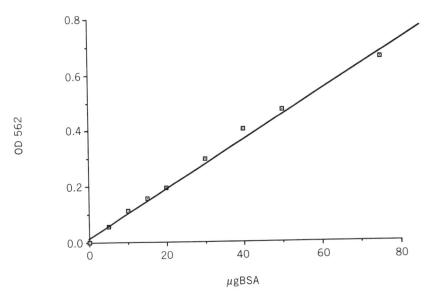

Figure 3.4. BCA Assay Standard Curve.

E. Comments

- To speed the development with the BCA assay, the protein assay tube can be incubated in a microwave oven for 20 seconds or less (Akins & Tuan, 1992).

- After cooling sample to room temperature, absorbance continues to increase at about 2.3%/10 min (Pierce BCA Handbook).

- The BCA microassay has a detection sensitivity of 0.5 - 10 µg/ml and requires more concentrated reagents and a 60 minute development time at 60°C (Pierce BCA Handbook).

- Samples containing lipids have spuriously high absorbances with the BCA assay, even relative to the Lowry assay (Kessler and Fanestil, 1986).

- Variations in BCA color development in buffers with sulfhydryl reagents and detergents can be found in Hill and Straka, 1988.

- Application of the deoxycholate-trichloroacetic acid (DOC-TCA) precipitation protocol (see Section III.D.) to the protein sample will remove most interfering substances.

V. Briefly Noted: Dot Filter Binding Assay

A. Summary

Dot filter binding (Coluccio and Bretscher, 1987) is a rapid way to obtain a preliminary determination of protein content in a large number of samples.   This assay is especially recommended for screening fractions from density gradient centrifugation or column chromatography. The information obtained from this initial screening can be used as a guide for pooling fractions from a protein peak and will reduce the number of samples which require more quantitative protein concentration determination.

B. Protocol

- **Materials**

    1. Whatman 3MM paper
    2. 10% Trichloroacetic Acid (TCA)
    3. Coomassie Gel Stain (see Chapter 5)
    4. Coomassie Gel Destain (see Chapter 5)

- **Assay**

    1. Using a pencil, prepare a 1 cm x 1 cm grid on 3MM paper.

    2. Apply 3 µl of each sample in the center of the grid square assigned to the sample.

    3. Allow filter paper to dry well, about 15 minutes.

    4. Soak filter paper thoroughly in 10% TCA for 30 seconds (TCA solution may be reused).

    5. Rinse paper briefly with water.

    6. Stain filter paper in Coomassie Gel Stain for 30 seconds (stain may be reused).

    7. Rinse paper briefly with water.

    8. Destain filter paper in Coomassie Gel Destain. Destaining should be complete within a few minutes.

    9. Identify fractions containing significant amounts of protein by the appearance of blue color in the corresponding squares.

VI. Interference Table

Highest Acceptable Concentrations

Compound	A_{280}	Bradford	Lowry	BCA
Buffers				
Acetate	0.1 M	0.6 M[12]	-	0.2 M[12]
Cacodylate-Tris	-	0.1M[9]	-	-
Glycine	1 M[12]	0.1 M[1]	2.5 mM[5]	1 M, pH11[6]
HEPES	-	100 mM[1]	2.5 μM[5]	100 mM[6]
MES	-	700 mM[1]	25 μM[5]	50 mM[7]
MOPS	-	200 mM[1]	25 μM[5]	50 mM[7]
Na$^+$-citrate	5%[12]	50 mM[1]	2.5 mM[5]	<1 mM[12]
PIPES	OK	500 mM[1]	5 μM[5]	50 mM[7]
Potassium phosphate	OK	1 M[1]	0.03 M[5]	-
Sodium phosphate	OK	1 M[1]	0.25 M[5]	0.1 M[6]
Sodium acetate	0.1 M[12]	0.6 M[12]	-	0.2 M, pH5.5[6]
TES	-	-	1 mM[5]	50 mM[7]
Tris	0.5 M[12]	2 M[1]	250 μM[5]	0.1 M[6]
Salts				
Ammonium sulfate	>50%[12]	1 M[1]	28 mM[5]	interferes[8]
NaCl	>1 M[12]	1 M[1]	30 mM[5]	1 M[12]
Urea	>1 M[12]	6 M[1]	200 mM[5]	3 M[6]
Detergents and Denaturants				
Brij 35	1%[12]	interferes[2]	-	1%[6]
CHAPS	10%[12]	1%[1]	1mM[11]	1%[6]
Guanidine-HCl	-	OK[1]	-	4M[8]
Na-deoxycholate	0.3%[12]	0.25%[1]	0.0625%[5]	-
Nonidet P-40	interferes	interferes[2]	-	1%[6]
Octyl glucoside	10%[12]	2%[2]	interferes[13]	1%[6]
SDS	OK	0.1%[1]	1.25%[5]	1%[6]
Sodium cholate	-	interferes[2]	-	-
Triton X-100	0.02%[12]	0.1%[1]	0.25%[5]	1%[6]
Tween-20	0.3%[12]	interferes[2]	-	-

(*continued*)

VI. Interference Table (continued)

Highest Acceptable Concentrations

Compound	A_{280}	Bradford	Lowry	BCA
Sugars				
Glucose	OK	OK[1]	30 mM[5]	10 mM[6]
Sucrose	2 M[12]	1 M[12]	10 mM[5]	1 M[6]
Chelators				
EDTA	30 mM[12]	100 mM[1]	125 µM[5]	10 mM[6]
EGTA	OK	0.05 M[1]	interferes	-
Reducing Agents				
2-mercaptoethanol	10 mM[12]	1 M[1]	1.75 mM[5]	50 µM[10]
DTT	interferes in oxidized state	1 M[1]	50 µM[5]	1 mM[6]
Alcohols, Polar Compounds				
Acetone	interferes	OK[1]	1.25%[5]	-
DMSO	20%[12]	-	6.2%[5]	5%[12]
Ethanol	OK	OK[1]	12.5%[5]	-
Ethylene glycol	OK	-	0.25%[5]	-
Glycerol	40%[12]	99%[1]	25%[5]	10%[6]
Methanol	OK	OK[1]	-	-

(*continued*)

VI. Interference Table (continued)

Highest Acceptable Concentrations

Compound	A_{280}	Bradford	Lowry	BCA
Miscellaneous				
Acrylamide	-	-	1.25 mg/ml[5]	-
DNA	interferes	1 mg/ml[1]	190 µg/ml[5]	-
Lipids	-		25µM[5]	interferes[9]
MgCl$_2$	OK	1 M[1]	-	-
RNA	interferes	0.3 mg/ml[1]	-	-
ATP	interferes	1 mM[1]	-	-
Ampholytes	-	1%[3,4]	interferes[5]	-
TCA-neutralized	10%[12]	-	12.5 mg/ml[5]	-
HCl	>1 M[12]	0.1 M[12]	-	0.1 N[6]
NaOH	>1 M[12]	0.1 M[12]	-	0.1 N[6]
NAD	interferes	1 mM[1]	-	-
Phenol	interferes	5%[1]	-	-
Amino acids	aromatics interfere	OK[1]	-	-
Polypeptides<3kd	interferes	OK[1]	-	-

- A dash (-) indicates that this compound has not been tested.
- "OK" indicates that this compound was used successfully but no concentration was indicated.
- Note that there may be some variation in permissible amounts due to differences in the assay protocol.

1 - Bio-Rad Bulletin 1069. 1987.   8 - Smith et al., 1985.
2 - Fanger, 1987.   9 - Kessler and Fanestil, 1986.
3 - Read and Northcote, 1981.   10 - Hill and Straka, 1988.
4 - Spector, 1978.   11 - Boehringer Mannheim Catalog
5 - Peterson, 1983.   12 - Stoscheck, 1990
6 - Pierce Bulletin 23225, 1984.   13 - Harris, 1989
7 - Kaushal and Barnes, 1986.

VII. Suppliers

3

Bicinchoninic Acid Assay: Fluka, Pierce (kit), Sigma

Bradford Assay: BioRad (kit), Fluka, Pierce (kit), Serva, Sigma

Dot Filter Binding Assay: Sigma, Whatman

Lowry Assay: Fluka, Merck, Sigma

VIII. References

A_{280}

Schleif, R.F. and P.C. Wensink. 1981. Practical Methods in Molecular Biology. New York, Springer-Verlag. p74.

Warburg, O. and W. Christian. 1941. Biochem. Z. 310: 384-421. Isolierung und Kristallisation des Gaerungsferments Enolase.

Wetlaufer, D.B. 1962. Adv. Prot. Chem. 17: 303-390. Ultraviolet Spectra of Proteins and Amino Acids

Whitaker, J.R. and P.E. Granum. 1980. Anal. Biochem. 109: 156-159. An Absolute Method for Protein Determination Based on Difference in Absorbance at 235 and 280 nm.

Bradford

Asryants, R.A., I.V. Duszenkova, and N.K. Nagradova. 1985. Anal. Biochem. 151: 571-574. Determination of Sepharose-Bound Protein with Coomassie Brilliant Blue G-250.

Bearden, J.C. 1978. Biochem. Biophys. Acta 533: 525-529. Quantitation of Submicrogram Quantities of Protein by an Improved Protein-Dye Binding Assay.

Bio-Rad Protein Assay. 1987. Bio-Rad Technical Bulletin 1069, Bio-Rad Laboratories.

Bradford, M.M. 1976. Anal. Biochem. 72: 248-254. A Rapid and Sensitive Method for the Quantitation of Microgram Quantities of Protein Utilizing the Principle of Protein-Dye Binding.

Fanger, B.O. 1987. Anal. Biochem. 162: 11-17. Adaptation of the Bradford Protein Binding Assay to Membrane-Bound Proteins by Solubilizing in Glucopyranoside Detergents.

Harris, E.L.V. 1989. Initial Planning. in Protein Purification Methods: A Practical Approach. Harris, E.L.V. & S. Angal, eds. pp.1-66. IRL Press, Oxford.

Peterson, G.L. 1983. Meth. Enzymol. 91: 95-121. Determination of Total Protein.

Pierce, J. and C.H. Suelter. 1977. Anal. Biochem. 81: 478-480. An Evaluation of the Coomassie Brilliant Blue G-250 Dye-Binding Method for Quantitative Protein Determination.

Read, S.M. and D.H. Northcote. 1981. Anal. Biochem. 116: 53-64. Minimization of Variation in the Response to Different Proteins of the Coomassie Blue G Dye-Binding Assay for Protein.

Spector, T. 1978. Anal. Biochem. 86: 142-146. Refinement of the Coomassie Blue Method of Protein Quantitation.

Van Kley, H. and S.M. Hale. 1977. Anal. Biochem. 81: 485-487. Assay for Protein by Dye Binding.

Lowry

Lowry, O.H., N.J. Rosebrough, A.L. Farr and R.J. Randall. 1951. J. Biol. Chem. 193: 265-275. Protein Measurement with the Folin Phenol Reagent.

Markwell, M.A.K., S.M. Haas, N.E. Tolbert and L.L. Bieber. 1981. Meth. Enzymol. 72: 296-303. Protein Determination in Membrane and Lipoprotein Samples.

Peterson, G.L. 1977. Anal. Biochem. 83: 346-356. A Simplification of the Protein Assay Method of Lowry et al. Which is More Generally Applicable.

Peterson, G.L. 1979. Anal. Biochem. 100: 201-220. Review of the Folin Phenol Protein Quantitation Method of Lowry.

Peterson, G.L. 1983. Meth. Enzymol. 91: 95-121. Determination of Total Protein.

Scopes, R.K. 1982. Protein Purification: Principles and Practice. New York, Springer-Verlag. pp. 240, 265-266.

BCA

Akins, R.E. and R.S. Tuan. 1992. BioTechniques 12: 496-499. Measurement of Protein in 20 Seconds Using a Microwave BCA Assay.

Hill, H.D. and J.G. Straka. 1988. Anal. Biochem. 170: 203-208. Protein Determination Using Bicinchoninic Acid in the Presence of Sulfhydryl Reagents.

Kaushal, V. and L.D. Barnes. 1986. Anal. Biochem. 157: 291-294. Effect of Zwitterionic Buffers on Measurement of Small Masses of Protein with Bicinchoninic Acid.

Kessler, R.J. and D.D. Fanestil. 1986. Anal. Biochem. 159: 138-142. Interference by Lipids in the Determination of Protein Using Bicinchoninic Acid.

Pierce Chemical Company (1984) BCA Protein Assay Reagent, Technical Bulletin 23225, Rockford, Ill.

Smith, P.K, R.I. Krohn, G.T. Hermanson, A.K. Mallia, F.H. Gartner, M.D. Provenzano, E.K. Fujimoto, N.M. Goeke, B.J. Olson and D.C. Klenk. 1985. Anal. Biochem. 150: 76-85. Measurement of Protein Using Bicinchoninic Acid.

Stoscheck, C.M. 1990. Meth. Enzymol. 182: 50-68. Quantitation of Protein.

Dot Filter Binding Assay

Coluccio, L.M. and A. Bretscher. 1987. J. Cell Biol. 105: 325-333. Calcium-regulated Cooperative Binding of the Microvillar 110K-Calmodulin Complex to F-Actin.

Chapter 4

Concentrating Protein Solutions

4

I. Analytical Methods
    A. Trichloroacetic Acid (TCA) Precipitation
    B. Acetone Precipitation
    C. Immunoprecipitation

II. Preparative Methods
    A. Ammonium Sulfate Precipitation ("Salting Out")
    B. Organic Solvent Precipitation
    C. Polyethylene Glycol (PEG) Precipitation
    D. Ultrafiltration
    E. Dialysis
    F. Briefly Noted: Ion Exchange Chromatography and Lyophilization

III. Suppliers

IV. References

I. Analytical Methods

The first two methods for protein concentration, acid and organic precipitations, result in protein denaturation. The reduced solubility of denatured proteins allows for their recovery in a pellet following centrifugation.

A. Trichloroacetic Acid (TCA) Precipitation

- **Equipment**

    1. Eppendorf tubes
    2. Eppendorf centrifuge
    3. Vortex mixer

- **Reagents**

    1. Trichloroacetic Acid (TCA), 100%
    2. NaOH, 0.1 N (400 mg NaOH in 100 ml H_2O)

- **Protocol** (Hames, 1981)

    1. To a 1 ml sample containing at least 5 µg of protein, add 100 µl of 100% trichloroacetic acid (TCA) and vortex.

    2. Allow protein to precipitate 30 minutes on ice (or 15 minutes in the freezer).

    3. Spin for 5 minutes at 10,000 x g.

    4. Remove supernatant by decanting immediately and aspirating remaining liquid with a pipette.

    5. Resuspend in 50 - 100 µl 0.1 N NaOH and vortex.

• **Comments**

1. Minimum protein concentration required for TCA precipitation is 5 μg/ml.

2. It is sometimes useful to wash the TCA pellet with 10% TCA or with ethanol-ether (1:1) to remove the TCA (Hames, 1981).

3. Instead of resuspending the TCA precipitate in NaOH (step 5), it may be more practical to carry out an ethanol-ether (1:1) wash and resuspend the pellet in a buffered solution.

4. Deoxycholate-trichloroacetic acid (DOC-TCA) precipitation is a common variation of the TCA precipitation which may extend the precipitation range of protein concentration to below 1 μg/ml (Peterson, 1983).
   a. To a 1 ml sample, add 0.1 ml 0.15% DOC.
   b. Vortex and incubate at room temperature for 10 minutes.
   c. Add 50 μl 100% TCA, vortex and go to step 3 (Peterson, 1977).

5. DOC-TCA precipitation may not work in the presence of SDS (Cabib and Polacheck, 1984).

B. Acetone Precipitation

- **Equipment**

    1. Eppendorf tubes
    2. Eppendorf centrifuge
    3. Vortex mixer

- **Reagents:** Acetone (or other organic solvent such as ethanol or methanol)

- **Protocol** (Hames, 1981)

    1. Add 1 ml of cold acetone (-20°C) to 200 μl of sample solution and vortex.

    2. Incubate at -20°C for 10 minutes.

    3. Centrifuge for 5 minutes in an Eppendorf centrifuge.

    4. Remove supernatant and air dry pellet.

    5. Resuspend pellet in 1 - 2 pellet volumes of buffer.

- **Comments**

    1. Protein pellet can be washed by repeated acetone precipitation.

    2. Quantitative precipitation of <1 μg of protein can be achieved by extending the -20°C incubation to >2 hours and spinning for 10 minutes at 27,000 x g (Sargent, 1987).

    3. Neutral salts increase protein solubility, while divalent cations reduce solubility in organic solutions (Kaufman, 1971).

C. Immunoprecipitation

A protein-specific antibody may permit quantitative isolation of the protein of interest by immunoprecipitation (Lerner and Steitz, 1979; see also Chapter 11, Section III.A.). This procedure consists of three steps. First, the specific antibody is added to the cell extract. In order to provide the mass necessary for precipitation from solution, chemically fixed *Staphylococcus aureus* bacteria are then added. These bacteria form complexes with the antibody via protein A, a bacterial protein which has a high affinity for the F_c portion of immunoglobulins. Alternatively, purified protein A coupled to beads of Sepharose offers a solid matrix for removing the antibody-antigen complex from the cell extract. Finally, a thorough wash of the pellet removes unprecipitated material.

Staphylococcus aureus cells should be washed prior to use for immunoprecipitation in order to remove damaged or poorly fixed cells (Kessler, 1981). If protein A-Sepharose beads are used, they should be swelled prior to use according to the manufacturer's instructions.

- **Equipment**
    1. Eppendorf tubes
    2. Eppendorf centrifuge
    3. Vortex mixer

- **Reagents**
    1. Protein-specific antibody
    2. Fixed *Staphylococcus aureus* cells or Protein A-Sepharose Beads
    3. Tris
    4. NaCl
    5. EDTA
    6. NaN_3
    7. Triton X-100

    Buffer A
    50 mM Tris-HCl (pH 7.4)
    150 mM NaCl
    5 mM EDTA
    0.02% NaN_3

4

- **Washing *S. aureus* cells**

  1. Within 24 hours of use, centrifuge the necessary volume of cells for 3 minutes in an Eppendorf centrifuge or at 3000 x *g* for 15 minutes.

  2. Resuspend in original volume with Buffer A containing 0.5% Triton X-100.

  3. Centrifuge as above.

  4. Resuspend in original volume with Buffer A containing 0.05% Triton X-100.

  5. Centrifuge as above.

  6. Resuspend in original volume with Buffer A.

  These washes can be carried out during step 2 of the immunoprecipitation protocol (see below).

- **Immunoprecipitation Protocol** (Kessler, 1981)

  1. Add a molar excess of specific antibody to the protein solution containing the antigen of interest. Typically, 10 - 25 μg of purified antibody are used in a 1 ml precipitation mix.

  2. Vortex and incubate precipitation mix on ice for at least 1 hour (2 - 3 hours for hybridoma supernatants).

  3. Add a sufficient quantity of *Staphylococcus aureus* cells or protein A-Sepharose beads to bind all immunoglobulin molecules in the precipitation mix. Calbiochem Standardized Pansorbin Cells provide an immunoglobulin binding capacity of 2.0 mg per ml of cells. Bio-Rad Affi-Gel protein A binds 6 - 7 mg of IgG per ml of gel.

  4. Vortex and incubate on ice for 15 - 60 minutes.

5. Centrifuge immunoprecipitate in an Eppendorf centrifuge for 15 seconds or spin at 3000 x g for 10 minutes at 4°C.

6. Wash immunoprecipitate 3 - 5 times in Buffer A containing 0.05% Triton X-100, taking care to resuspend the pellet completely with each wash.

   a. Resuspend pellet in 100 µl washing buffer by vortexing.
   b. Centrifuge for 15 seconds in an Eppendorf centrifuge.
   c. Repeat steps **a** and **b** 3 - 5 times.

7. Pellets can be resuspended in 100 µl of sample buffer (see Chapter 5, section I.C.) or isoelectric focusing buffer (see Chapter 7, Section I.G.). A quarter to a half of the pellet is often sufficient for detection on a gel, although this depends on the starting protein concentration. Heating to 100°C will be necessary for efficient separation of the protein complex.

- **Comments**

   1. The antibody solution should be ultracentrifuged (160,000 x g for 30 minutes in an airfuge or 400,000 x g for 10 min in a table-top ultracentrifuge) prior to use in order to remove aggregates (Kessler, 1981). Repeated freeze-thawing of antibody solutions will lead to increased aggregation, so the antibody should be stored in aliquots. Antibody may be incubated at 56°C for 30 minutes prior to use in order to inactivate proteases (Scheidtmann, 1989).

   2. *Staphylococcus aureus* cells may be cultured and prepared for immunoprecipitation (see Kessler, 1981) or purchased already formalin-fixed and heat-treated. Prepared cells may be stored as aliquots at -20°C.

   3. Nonspecific *S. aureus* binding can be reduced by preabsorbing washed *S. aureus* cells with antigen extract (Firestone & Winguth, 1990).

   4. The specificity of the immunoprecipitate may be improved by changing the pH, salt concentration or detergent utilized in the washing buffer.

5. *Staphylococcus aureus* has different binding affinities for different immunoglobulins (Schantz, 1983). Good binding is found with rabbit, human, and guinea pig immunoglobulins. Differential binding with different immunoglobulin subclasses is seen with the following animals:

	Good	Poor
Mouse	IgG_{2a}, IgG_{2b}, IgG_3	IgG_1
Rat	IgG_1, IgG_{2c}	IgG_{2a}, IgG_{2b}
Goat	IgG_2	IgG_1
Sheep	IgG_2	IgG_1

6. For mouse and goat antibodies, maximal binding is found above pH 8 and 9, respectively. To compensate for poor mouse immunoglobulin subclass 1 binding, for example, it is possible to preincubate the *S. aureus* cells with anti-mouse IgG antibody.

7. Protein G is isolated from a *Streptococcus* strain and is similar to protein A in its properties of binding antibodies. Protein G possesses better binding properties than protein A to antibodies from various species (see Chapter 11). A protein G derivative with the albumin binding domain deleted has been coupled to Sepharose and is available commercially (Pierce, Pharmacia).

8. Immunoprecipitates may require urea in sample buffer to aid in solubilization.

9. Conditions used for antigen release following immunoprecipitation include 3.5 M $MgCl_2$, 0.1 M NaCitrate (pH 3.0), 3 M ammonium or potassium thiocyanate, saturated guanidine hydrochloride, 0.2 M lithium diiodosalicylate, and 2-mercaptoethanol.

10. This procedure is most often used for isolating radiolabeled proteins and is followed by autoradiography (Kessler, 1981). Alternatively, the immunoprecipitate may be detected by immunoblotting; however, controls must be included to rule out cross-reactivity with *S. aureus* proteins.

11. Suggestions for running proper controls and reducing background precipitation can be found in Harlow and Lane (1988, pp. 465, 469-470).

II. Preparative Methods

Methods for concentrating proteins which permit retention of protein activity are discussed in the following portion of this chapter.

4

A. Ammonium Sulfate Precipitation ("Salting Out")

When high concentrations of salt are present, proteins tend to aggregate and precipitate out of solution. This technique is referred to as "salting out". Since different proteins precipitate at different salt concentrations, salting out is often used during protein purification. It is important to remember that factors such as pH, temperature and protein purity play important roles in determining the salting out point of a particular protein (theory: Scopes, 1981).

Ammonium sulfate is the salt of choice because it combines many useful features such as salting out effectiveness, pH versatility, high solubility, low heat of solution and low price (Scopes, 1981).

Ammonium sulfate concentrations are generally expressed in percent saturation, and a simple equation for calculation of grams of ammonium sulfate needed to make an X% solution starting from X_0% is:

$$g = \frac{515 (X - X_0)}{100 - 0.27X} \quad \text{(for a 1 liter solution at } 0°C\text{);}$$
$$\text{see also Appendix 3}$$

Since most proteins will precipitate at 55% ammonium sulfate, a good value for obtaining maximum protein precipitation is 85%. For a 100 ml solution containing no ammonium sulfate at the start, the following protocol is recommended:

4

- **Equipment**

    1. Beakers
    2. Magnetic stir plate and stir bar
    3. Centrifuge and rotor

- **Reagents**

    1. Ammonium sulfate
    2. Buffer in which to resuspend pellet

- **Protocol**

    1. Place beaker of the protein solution in a cooling bath on top of a magnetic stir plate. This can be accomplished by placing the beaker within another beaker containing a water - ice slurry.

    2. While agitating gently on a magnetic stirrer, slowly add 56.8 g ammonium sulfate. Add salt more slowly as final saturation is approached. This step should be completed in 5 - 10 minutes.

    3. Continue stirring for 10 - 30 minutes after all salt has been added.

    4. Spin at 10,000 x g for 10 minutes or at 3000 x g for 30 minutes.

    5. Decant supernatant and resuspend precipitate in 1 - 2 pellet volumes of buffer. Any insoluble material remaining is probably denatured protein and should be removed by centrifugation.

    6. Ammonium sulfate can be removed by dialysis, ultrafiltration, or a desalting column.

- **Comments**

    1. Stirring must be regular and gentle. Stirring too rapidly will cause protein denaturation as evidenced by foaming. It is also important to use a magnetic stirrer which does not generate a significant amount of heat while stirring.

    2. While most proteins precipitate in the first 20 minutes after the salt is dissolved, some require several hours to completely precipitate.

    3. Precipitation should be carried out in a buffer of at least 50 mM in order to compensate for a slight acidification upon dissolving ammonium sulfate.

    4. The buffer should contain a chelating agent such as EDTA to remove possible traces of heavy metal cations in the ammonium sulfate which might be detrimental to the protein of interest.

    5. To ensure maximal precipitation, it is best to start with a protein concentration of at least 1 mg/ml.

    6. Ammonium sulfate precipitation is often a good way of stabilizing proteins for storage (Scopes, 1981; see Chapter 1, Section VI.C.).

    7. Few proteins precipitate below 24% ammonium sulfate while most do above 80%. Solubility differences in ammonium sulfate are frequently exploited to separate proteins in the early stages of purification protocols. Frequently, ammonium sulfate precipitation results in removal of RNA and DNA (Schleif and Wensink, 1981).

    8. A protein's solubility may be reduced at its isoelectric point where electrostatic interactions can lead to protein aggregation and precipitation (Scopes, 1981). A lower ammonium sulfate concentration will be required to precipitate a protein at its isoelectric point.

    9. The effectiveness of ammonium sulfate in precipitating proteins is often exploited in crystallization experiments (see Chapter 12).

B. Organic Solvent Precipitation

Organic solvents such as acetone and ethanol have effects similar to high levels of salt when added to protein solutions; that is, they lower the protein solubility (for further explanation, see Chapter 12, Section II.A.). Proteins are more easily denatured in organic solvents at temperatures above 10°C, so special care must be taken to work with chilled solutions and centrifuge rotors (0 - 4°C is satisfactory). Ionic strengths between 0.05 and 0.2 M are recommended (Kaufman, 1971; Scopes, 1982).

Concentrations of organic solvents are generally calculated in volume percentages assuming that volumes are additive, which is usually not strictly the case. For example, in the conventional procedure to bring a 100 ml aqueous solution to 50% (v/v) ethanol, one would add 100 ml of ethanol even though the final volume only measures 192 ml. Most proteins larger than 15 kd precipitate with 50% organic solvent.

- **Equipment**
    1. Beaker
    2. Magnetic stir plate and stir bar
    3. Centrifuge and rotor

- **Reagents**
    1. Acetone or ethanol
    2. Buffer in which to resuspend pellet

- **Protocol**

   1. To 100 ml of protein solution at 4°C, slowly add 100 ml acetone or ethanol (chilled to -20°C) with continuous gentle stirring on ice.

   2. Continue stirring on ice for 10 - 15 minutes.

   3. Centrifuge in a cold rotor for 10 minutes at 10,000 x g.

   4. Decant supernatant.

   5. Gently resuspend pellet in two pellet volumes of cold buffer. Particles which are difficult to dissolve are likely to be denatured proteins and should be removed by filtration or centrifugation.

- **Comments**

   1. Acidic proteins may be more easily precipitated as complexes with divalent cations such as magnesium.

   2. Larger proteins as well as more hydrophilic ones tend to come out of solution at lower organic solvent concentrations.

   3. A pH closer to the protein's isoelectric point reduces its solubility in organic solvents.

C. Poly(ethylene glycol) (PEG) Precipitation

Poly(ethylene glycol), a nonionic water-soluble polymer, causes little protein denaturation while inducing precipitation at a discrete PEG concentration for a given protein (Ingham, 1984). This property, along with its low heat of solution and short equilibration time for precipitation, makes it a useful reagent for protein fractionation as well as crystallization (see Chapter 12). Maximum protein precipitation is generally achieved with a final PEG concentration of 30%.

- **Equipment**

    1. Beaker
    2. Magnetic stir plate and stir bar
    3. Centrifuge and rotor

- **Reagents:** Poly(ethylene glycol) 6000

- **Protocol**

    1. To a 100 ml protein solution, slowly add 150 ml 50% PEG-6000 (w/v, in distilled water) while stirring gently.

    2. Continue stirring for 30 - 60 minutes to allow precipitation to come to completion.

    3. Centrifuge 10 minutes at 10,000 x g.

    4. Decant supernatant.

    5. Resuspend pellet in 1 - 2 pellet volumes of buffer.

- **Comments**

    1. The solubility of a protein is reduced near its isoelectric point.

    2. To remove PEG from the precipitate, an ammonium sulfate precipitation is suggested. Alternatively or in addition, ion exchange chromatography has been used (Fried and Chun, 1971). Gel filtration columns, however, have been reported to behave anomalously in the presence of elevated levels of PEG.

D. Ultrafiltration

Concentration of protein by ultrafiltration proceeds by forcing the liquid in a protein solution through a membrane which retains the protein of interest. The majority of laboratory ultrafiltrations involve relatively small volumes (several milliliters) of protein which are most easily handled by using centrifugal force as the means of forcing liquid through the membrane. Thus, our description will be limited to the Amicon Centricon system as an example of ultrafiltration (Amicon Publication No. I-259C, 1986). This method is less likely to cause protein denaturation than the precipitative methods described above.

The Centricon Microconcentrator has a starting capacity of 2 ml. A two ml sample of 2 mg/ml bovine serum albumin (BSA) may be concentrated to less than 50 μl by centrifuging for as little as 30 minutes. However, centrifugation times can increase significantly depending on the type of ultracentrifugation membrane used, the composition of the buffer, and the type and concentration of protein (see Comments below). By adding a new buffer to the concentrated sample and recentrifuging, buffer changes and desalting can be easily achieved (Penefsky, 1977; Christopherson, 1983). Four models are available, Centricon-3, 10, 30 and Centricon-100, with 3, 10, 30, and 100 kd average molecular weight cutoffs.

- **Equipment**

    1. Amicon Centricon Microconcentrator
    2. Centrifuge and rotor

- **Reagents:** None

- **Protocol**

    1. Assemble the Centricon Microconcentrator and add up to 2.0 ml of sample to the sample reservoir (Fig. 4.1).

    2. Centrifuge at 5000 x *g* for 30 minutes.

    3. Remove filtrate cup and apply retentate cup to the sample reservoir.

    4. Invert the Centricon unit and centrifuge at 1000 x *g* for 2 minutes.

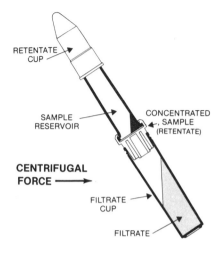

Figure 4.1. Concentration of a protein solution using the Amicon Centricon system.

5. The concentrated sample will be recovered in the retentate cup. A small amount of buffer may be used to wash the filter for more complete protein recovery.

• **Comments**

1. Centrifuge times may be decreased depending on the characteristics of the sample. Viscous solutions containing glycerol or high protein concentrations, for example, require much longer centrifugation times.

2. Recovery of proteins is typically greater than 90%.

3. The protein of interest should be 30 - 50% larger than the stated cutoff to ensure that it will be retained.

4. Different average pore sizes in different membranes may permit a slight enrichment of the protein by selective retention on the basis of protein size.

5. The Centricon Microconcentrator is designed to be used in a fixed-angle rotor. Depending on the angle of the rotor, the sample will reach a deadstop volume of 25 - 40 µl beyond which no further concentration will occur. This prevents the sample from being filtered to dryness.

4

6. Other centrifuge-based ultrafiltration systems include the Amicon Centriflo (7 ml capacity) and the Centriprep (15 ml capacity) Concentrators.

7. For large volume applications, pressurized ultrafiltration systems are much more time-efficient. Stirred cell systems from Amicon are suitable for volumes of up to several hundred milliliters, while larger volumes are handled with spiral, hollow fiber, or cassette-type membrane cartridges. The reader is referred to the Amicon catalog for further details.

8. Penefsky (1977) describes a simple centrifuge column which may be used with gel filtration resin for salt removal (see also Chapter 10, Section II).

E. Dialysis

Dialysis is typically used for changing the buffering solution of a protein, but it can also be used as a method for concentrating protein solutions if carried out in a vacuum or a hygroscopic environment (e.g. PEG, Sephadex). The protein solution is contained within a membrane whose pore size prevents the protein from escaping and which permits solute exchange with either air at reduced pressure or a surrounding solution.

Except for small volumes, this diffusion-based method is time-consuming, and ultrafiltration is often preferable for preparative protein concentration and buffer exchange. However, in some cases, time is not a limiting factor and dialysis may be preferred (theory: McPhie, 1971). Pierce now offers premade dialysis units which require no preparation time and no specialized equipment.

- **Equipment**

    1. Beaker
    2. Magnetic stir plate and stir bar
    3. Dialysis tubing

- **Reagents**: Sephadex G-100 or G-200

- **Preparing Dialysis Tubing**

    Dialysis tubing contains chemical contaminants from the manufacturing process. To remove these, it is common to boil the tubing for at least 30 minutes in 10 mM sodium bicarbonate ($NaHCO_3$)/1mM EDTA. Some authors recommend more thorough treatments (see Richmond et al., 1985). Following the boiling step, the tubing should be washed extensively in distilled water and stored at 4°C in 1 mM EDTA to prevent microbial contamination. Prepare new dialysis tubing every six months (Schleif and Wensink, 1981). Spectrum offers a line of dialysis membranes (Spectra/Por 7) made from regenerated cellulose which are pretreated to remove heavy metal and sulfide contaminants.

- **Protocol**

    1. Make two tight knots at one end of the tubing.

    2. Using a pipette or funnel (Fig. 4.2), deliver protein solution into the dialysis tubing.

    3. Tie a double knot at the other end of the tubing and place the closed dialysis bag in >10 volumes of dialysis buffer. The buffer should be stirred gently with a magnetic stir bar to improve solute exchange. Equilibrium occurs after several hours of dialysis, and the dialysis buffer may have to be changed several times until certain buffer components are sufficiently diluted.

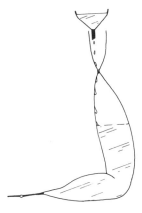

Figure 4.2. Filling dialysis tubing with a protein solution.

- **Dialysis Buffer**

    For the purposes of protein concentration, the dialysis bag may be incubated in a buffer containing 20% poly(ethylene glycol) (PEG 20,000) or 5% ethylene glycol. However, use of PEG may be accompanied by leaching of unwanted compounds into the dialysate which might require a further purification step. A more rapid method for concentrating a protein solution is to embed the dialysis sack in Sephadex G-100 or G-200 resin or Calbiochem Aquacides at 4°C. A fivefold concentration can be achieved in 3 hours by changing the resin surrounding the dialysis sack every half hour (Schleif and Wensink, 1981).

4

• **Comments**

1. Commercially available dialysis clips (Spectrum) are popular as an alternative to tying knots in the dialysis tubing.

2. If dialyzing against a buffer with a lower salt or organic solvent concentration than the dialysate, be aware that osmotic forces will cause water influx into the dialysis sack. Allow space for volume increases to avoid the risk of the membrane bursting.

3. Dialysis tubing is available in a variety of molecular weight cutoffs from 100 to 300,000 daltons (Spectrum Medical Industries).

4. A description of a simple microtechnique for dialyzing small volumes (10 μl to 0.6 ml) in an Eppendorf tube can be found in Overall (1987).

F. Briefly Noted: Ion Exchange Chromatography and Lyophilization

Ion Exchange Chromatography

4

Most proteins display an overall negative charge at pH 8 and thus are candidates for concentration on an anion exchange resin. The methods for preparing the ion exchange resin and carrying out column chromatography are described in Chapter 9, Section II.

Lyophilization

Lyophilization is a commonly used method for concentration and storage of protein solutions involving sublimation of liquid from the sample in the frozen state. The reader is referred to Everse and Stolzenbach (1971) for advice and to their lyophilizer instruction manual for operating instructions.

4

III. Suppliers

    Dialysis Membranes: Spectrum

    Fixed *Staphylococcus aureus* cells: Bethesda Research Laboratories, Calbiochem

    Microconcentrator: Amicon Centricon, Bio-Rad Unisep Cartridges

    Protein A - Sepharose Beads: Bio-Rad, Pierce

    Protein G - Sepharose Beads: Pharmacia, Pierce

    TCA: Fluka, Merck, Serva, Sigma

    Triton X-100: Sigma

    Ultrafiltration Stirred Cells: Amicon, Sartorius, Schleicher & Schuell

IV. References

Amicon Publication No. I-259C. 1986. Centricon Microconcentrators for Small-Volume Concentration.

Cabib, E. and I. Polacheck. 1984. Meth. Enzymol. 104: 415-416. Protein Assay for Dilute Solutions.

Christopherson, R.I. 1983. Meth. Enzymol. 91: 278-281. Desalting Protein Solutions in a Centrifuge Column.

Everse, J. and F.E. Stolzenbach. 1971. Meth. Enzymol. 22: 33-39. Lyophilization.

Firestone, G.L. and S.D. Winguth. 1990. Meth. Enzymol. 182: 688-700. Immunoprecipitation of Proteins.

Fried, M. and P.W. Chun. 1971. Meth. Enzymol. 22: 238-248. Water-Soluble Nonionic Polymers in Protein Purification.

Hames, B.D. 1981. in Gel Electrophoresis of Proteins. A Practical Approach. Hames, B.D. and D. Rickwood, eds. 290 pages. IRL Press Limited, London.

Harlow, E. and D. Lane. 1988. Antibodies: A Laboratory Manual. 726 pages. Cold Spring Harbor Laboratory, Cold Spring Harbor, New York.

Ingham, K.C. 1984. Meth. Enzymol. 104: 351-355. Protein Precipitation with Polyethylene Glycol.

Kaufman, S. 1971. Meth. Enzymol. 22: 233-238. Fractionation of Protein Mixtures with Organic Solvents.

Kessler, S.W. 1981. Meth. Enzymol. 73: 442-459. Use of Protein A-Bearing Staphylococci for the Immunoprecipitation and Isolation of Antigens from Cells.

Lerner, M.R. and J.A. Steitz. 1979. Proc. Nat. Acad. Sci. USA 76: 5495-5499. Antibodies to Small Nuclear RNAs Complexed with Proteins are Produced by Patients with Systemic Lupus Erythematosus.

McPhie, P. 1971. Meth. Enzymol. 22: 23-33. Dialysis.

Nelson, N. 1986. Meth. Enzymol. 118: 352-369. Subunit Structure and Biogenesis of ATP Synthase and Photosystem I Reaction Center.

Overall, C.M. 1987. Anal. Biochem. 165: 208-214. A Microtechnique for Dialysis of Small Volume Solutions with Quantitative Recoveries.

Penefsky, H.S. 1977. J. Biol. Chem. 252: 2891-2899. Reversible Binding of P_i by Beef Heart Mitochondrial Adenosine Triphosphatase.

4

Peterson, G.L. 1977. Anal. Biochem. 83: 346-356. A Simplification of the Protein Assay Method of Lowry et al. Which is More Generally Applicable.

Pharmacia Fine Chemicals. Ion Exchange Chromatography: Principles and Methods. 71 pages.

Richmond, V.l., R. St. Denis and E. Cohen. 1985. Anal. Biochem. 145: 343-350. Treatment of Dialysis Membranes for Simultaneous Dialysis and Concentration.

Sargent, M.G. 1987. Anal. Biochem. 163: 476-481. Fiftyfold Amplification of the Lowry Protein Assay.

Shantz, E.M. 1983. PANSORBIN *Staphylococcus aureus* Cells: Review and Bibliography of the Immunological Applications of Fixed Protein A-Bearing *Staphylococcus aureus* Cells. Calbiochem booklet. 56 pages.

Scheidtmann, K.H. 1989. pp. 109-112 in Protein Structure: A Practical Approach. T.E. Creighton, ed. 355 pages. IRL Press, Oxford.

Schleif, R.F. and P.C. Wensink. 1981. Practical Methods in Molecular Biology. Springer-Verlag, New York.

Scopes, R.K. 1982. Protein Purification. Principles and Practice. Springer-Verlag, New York.

Chapter 5

Gel Electrophoresis under Denaturing Conditions

5

I.   SDS-Polyacrylamide Gel Electrophoresis (Linear Slab Gel)
    A. Introduction
    B. Equipment
    C. Pouring a Gel
    D. Preparing and Loading Samples
    E. Running a Gel
    F. Staining a Gel with Coomassie Blue
    G. Silver Staining a Gel
    H. Drying a Gel
    I.   General Discussion
    J.   Safety Notes

II. Gradient Gels
    A. Introduction
    B. Equipment
    C. Preparing Gels

III. SDS-Urea Gels
    A. Introduction
    B. Preparing Gels

IV. Other Methods
    A. Detection of Radiolabeled Samples
    B. Molecular Weight Determination
    C. Protein Quantitation (Densitometry)
    D. Eluting Protein Bands Following Electrophoresis

V.   Suppliers

VI. References

I. SDS-Polyacrylamide Gel Electrophoresis
   ## (Linear Slab Gel)

A. Introduction

Sodium dodecyl sulfate - polyacrylamide gel electrophoresis (SDS-PAGE) is a low-cost, reproducible, and rapid method for quantifying, comparing, and characterizing proteins. This method separates proteins based primarily on their molecular weights (Laemmli, 1970). SDS binds to hydrophobic portions of a protein, disrupting its folded structure and allowing it to exist stably in solution in an extended conformation. As a result, the length of the SDS-protein complex is proportional to its molecular weight. The ease of execution and wide application of SDS-PAGE have made it an important analytical technique in many fields of research.

Our description of SDS-PAGE will include protocols for linear slab gels, i.e., gels formed as slabs between two sheets of supporting glass. Slab gels have become more widely used than tube gels (formed with glass tube supports), since many samples can be run on the same gel, thereby providing uniformity during polymerization, staining, and destaining. For most analytical applications, the mini slab gel has gained considerable popularity due to the increased resolution and reduced amounts of time and materials required for electrophoresis. The experimental procedures and reagents in this volume have been calculated for use with a mini-gel system; however, all procedures should be readily adaptable to other systems. For certain applications, gradient gels are useful (see Section II). Pre-formed homogeneous and gradient slab gels can also be purchased from various suppliers (Novex, ISS, Pharmacia, Bio-Rad).

- Among the varied uses of this technique are:
    1. Analysis of protein purity
    2. Determination of protein molecular weight
    3. Verification of protein concentration
    4. Detection of proteolysis
    5. Identification of immunoprecipitated proteins
    6. First stage of immunoblotting (see Chapter 8)
    7. Detection of protein modification
    8. Separation and concentration of protein antigens for antibody production
    9. Separation of radioactively labeled proteins

- Theory: Blackshear, 1984

- Sensitivity of staining:
    1. Coomassie Blue: 0.1 - 1 µg per band (Smith, 1984)
    2. Silver Staining: 2 - 10 ng per band (Giulian et al., 1983)

- Optimal Resolution Ranges (adapted from Hames, 1981)

Acrylamide Percentage	Separating Resolution
15% Gel	15 to 45 kd
12.5% Gel	15 to 60 kd
10% Gel	18 to 75 kd
7.5% Gel	30 to 120 kd
5% Gel	60 to 212 kd

- Time Required:

    1. Individual Steps:
        Pouring Separating Gel, 60 minutes; Pouring Stacking Gel, 30 minutes; Loading Samples, 15 minutes; Electrophoresis, 45 minutes
        Staining
        Coomassie Staining   30 minutes (major bands)
        Silver Staining   3 hours

    2. Total Time:
        Coomassie Stained Gel: 3 hours for major bands to destain. Complete destaining may require 24 - 48 hours.
        Silver Stained Gel: 6 hours

B. Equipment

- Minigel apparatus
    We use the Bio-Rad Mini-Protean II apparatus as a model, although all protocols should be easily adaptable to other systems. We highly recommend the minigel systems due to the savings in material and time and also because they provide high resolution protein separation.
- Power supply (capacity 200 V, 500 mA)
- Boiling water bath or 100°C sand bath
- Eppendorf centrifuge (optional)
- Hamilton Syringes or disposable gel loading tips for micropipette
- Gel dryer and high vacuum pump or water pump (optional)
- Small glass or plastic container with lid (i.e. 12 x 16 x 3 cm)
- Eppendorf tubes
- Rocking or rotary shaker

C. Pouring a Gel

- **Reagents**
    1. Acrylamide, electrophoresis grade
    2. Bis-acrylamide (*N,N'*-methylenebisacrylamide)
    3. Tris (2-hydroxymethyl-2-methyl-1,3-propanediol)
    4. SDS (sodium dodecyl sulfate or sodium lauryl sulfate)
    5. TEMED (*N,N,N',N'*-tetramethylene-ethylenediamine)
    6. Ammonium persulfate
    7. 2-mercaptoethanol
    8. Glycerol
    9. Bromophenol blue
    10. Glycine
    11. Hydrochloric acid (HCl)
    12. Dithiothreitol (DTT)

- **Stock Solutions**

    1. **2 M Tris-HCl (pH 8.8)**, 100 ml
        a. Weigh out 24.2 g Tris base
        b. Add to 50 ml distilled water
        c. Add concentrated HCl slowly to pH 8.8 (about 4 ml)
           (allow solution to cool to room temperature, pH will
           increase)
        d. Add distilled water to a total volume of 100 ml

    2. **1 M Tris-HCl (pH 6.8)**, 100 ml
        a. Weigh out 12.1 g Tris base
        b. Add to 50 ml distilled water
        c. Add concentrated HCl slowly to pH 6.8 (about 8 ml)
           (allow solution to cool to room temperature, pH will
           increase)
        d. Add distilled water to a total volume of 100 ml

    3. **10% SDS** (w/v), 100 ml, store at room temperature
        a. Weigh out 10 g SDS
        b. Add distilled water to a total volume of 100 ml

    4. **50% glycerol** (v/v), 100 ml
        a. Pour 50 ml 100% glycerol
        b. Add 50 ml distilled water

    5. **1% bromophenol blue** (w/v), 10 ml
        a. Weigh out 100 mg bromophenol blue
        b. Bring to 10 ml with distilled water, stir until dissolved
           Filtration will remove aggregated dye.

- **Working Solutions**

   1. Solution A (Acrylamide Stock Solution), 100 ml
   30% (w/v) acrylamide, 0.8% (w/v) bis-acrylamide

   Caution: Unpolymerized acrylamide is a skin irritant and a neurotoxin. Always handle with gloves. See Safety Notes.

   a. 29.2 g acrylamide
   b. 0.8 g bis-acrylamide
   Add distilled water to make 100 ml and stir until completely dissolved. Work under hood and keep acrylamide solution covered with Parafilm until acrylamide powder is completely dissolved.
   Can be stored for months in the refrigerator.

   2. Solution B (4x Separating Gel Buffer), 100 ml

   a. 75 ml 2 M Tris-HCl (pH 8.8)   ---1.5 M
   b. 4 ml 10% SDS   ---0.4%
   c. 21 ml H_2O
   Stable for months in the refrigerator.

   3. Solution C (4x Stacking Gel Buffer), 100 ml

   a. 50 ml 1 M Tris-HCl (pH 6.8)   ---0.5 M
   b. 4 ml 10% SDS   ---0.4%
   c. 46 ml H_2O
   Stable for months in the refrigerator.

   4. 10% ammonium persulfate, 5 ml

   a. 0.5 g ammonium persulfate
   b. 5 ml H_2O
   Stable for months in a capped tube in the refrigerator.

   5. Electrophoresis Buffer, 1 liter

   a. 3 g Tris   ---25mM
   b. 14.4 g glycine   ---192 mM
   c. 1 g SDS   ---0.1%
   d. H_2O to make 1 liter
   pH should be approximately 8.3.
   Can also make a 10x stock solution.
   Stable indefinitely at room temperature.

6. 5x Sample Buffer, 10 ml

a. 0.6 ml 1 M Tris-HCl (pH 6.8)   ---60 mM
b. 5 ml 50% glycerol   ---25%
c. 2 ml 10% SDS   ---2%
d. 0.5 ml 2-mercaptoethanol   ---14.4 mM
e. 1 ml 1% bromophenol blue   ---0.1%
f. 0.9 ml H_2O

> Stable for weeks in the refrigerator or for months at -20°C.

- **Amounts of Working Solutions to Use**

1. Volumes necessary for pouring gels of different thicknesses (for two 6 x 8 cm gels)

Gel Thickness	Separating	Stacking
0.5 mm	5.6 ml	1.4 ml
0.75 mm	8.4 ml	2.1 ml
1.0 mm	11.2 ml	2.8 ml
1.5 mm	16.8 ml	4.2 ml

Always prepare with a moderate excess of gel solution.

2. Calculation for X% Separating Gel

Solution A $^x/_3$ ml

Solution B   2.5 ml

H_2O $(7.5-^x/_3)$ ml

10% Ammonium Persulfate   50 µl

TEMED   5 µl (10 µl if x<8%)

_____

Total Volume   10 ml

Do not prepare until following instructions in the next section, Pouring the Separating Gel.

- **Pouring the Separating Gel**

Example of Separating Gel Preparation

Two 8% Separating Gels (for two 6 cm x 8 cm x
0.75 mm gels, need 10 ml)

2.7 ml Solution A
2.5 ml Solution B
4.8 ml H_2O

---

50 µl 10% Ammonium Persulfate
5 µl TEMED

**Do not prepare until following numbered instructions
below.**

1. **Assemble gel sandwich** according to the manufacturer's
   instructions in the case of commercial apparatus (e.g., Bio-
   Rad Mini-Gel), or according to the usage of alternative
   systems. For Mini-Gel, be sure that the bottom of both gel
   plates and spacers are perfectly flush against a flat surface
   before tightening clamp assembly (Fig. 5.1). A slight
   misalignment will result in a leak.

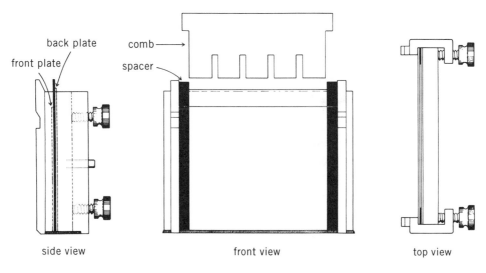

Figure 5.1. Bio-Rad Mini-Protean II apparatus.
Views of gel plate assembly.

5

2. **Combine Solutions A and B and water** in a small Erlenmeyer flask or a test tube. Acrylamide (in Solution A) is a neurotoxin, so plastic gloves should be worn at all times

3. **Add ammonium persulfate and TEMED, and mix** by swirling or inverting container gently (excessive aeration will interfere with polymerization). Work rapidly at this point because polymerization will be under way.

4. Carefully **introduce solution into gel sandwich** using a pipette. Pipet solution so that it descends along a spacer (Fig. 5.2). This minimizes the possibility of air bubbles becoming trapped within the gel.

5. When the appropriate amount of separating gel solution has been added (in the case of the Mini-Gel, about 1.5 cm from top of front plate or 0.5 cm below level where teeth of comb will reach, Fig. 5.3), gently **layer about 1 - 5 mm of water** on top of the separating gel solution. This keeps the gel surface flat.

6. **Allow gel to polymerize** (30 - 60 minutes).

When the gel has polymerized, a distinct interface will appear between the separating gel and the water, and the gel mold can be tilted to verify polymerization. It is a good idea to draw some of the unused separating gel solution into a Pasteur pipette immediately after pouring the gel. This serves as a check for polymerization.

If the gel leaks (and you have not yet covered it with water), it may be possible to recover the separating gel solution, reposition the plates and spacers, and repour the gel before complete polymerization has occurred.

Separating gels can be stored for up to a week at 4°C. Remove water, replace with Solution B diluted 1:3, and cover with plastic wrap.

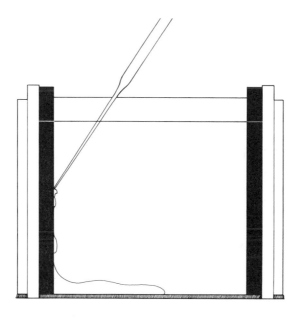

Figure 5.2. Introducing the separating gel solution into the gel sandwich.

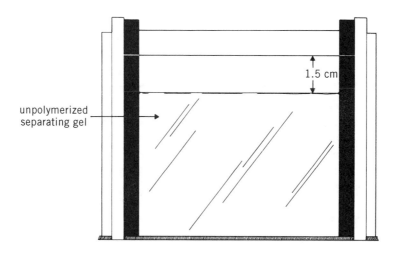

Figure 5.3. Separating gel prior to polymerization.

5

- **Pouring the Stacking Gel**

  Example of Standard Stacking Gel Preparation

  Two 5% Stacking Gels (for two 6 cm x 8 cm x 0.75 mm gels, need 4 ml)

  2.3 ml H_2O
  0.67 ml Solution A
  1.0 ml Solution C
  _____

  30 μl 10% Ammonium Persulfate
  5 μl TEMED

  Do not prepare until following the numbered instructions below.

  1. **Pour off water** covering the separating gel. The small droplets remaining will not disturb the stacking gel.

  2. **Combine Solutions A and C and water** in a small Erlenmeyer flask or a test tube.

  3. **Add ammonium persulfate and TEMED and mix** by gently swirling or inverting the container.

  4. **Pipet stacking gel solution onto separating gel** until solution reaches top of front plate (Fig. 5.4).

  5. **Carefully insert comb** into gel sandwich until bottom of teeth reach top of front plate (Figs. 5.5 and 5.6). Be sure no bubbles are trapped on ends of teeth. Tilting the comb at a slight angle is helpful for insertion without trapping air bubbles.

  6. **Allow stacking gel to polymerize** (about 30 minutes).

7. After stacking gel has polymerized, **remove comb** carefully, making sure not to tear the well ears.

8. Place gel into electrophoresis chamber. If using the Mini-Gel system, **attach both gels to electrode assembly before inserting into electrophoresis tank.**

9. **Add electrophoresis buffer to inner and outer reservoir**, making sure that both top and bottom of gel are immersed in buffer (Fig. 5.7).

5

Check wells for trapped air bubbles and damaged well ears. Distorted well ears can be repositioned using a Hamilton syringe.

Air bubbles clinging to bottom of gel should be removed to insure even current flow.

It is useful to rinse wells with electrophoresis buffer prior to loading in order to remove unpolymerized acrylamide and any contaminants.

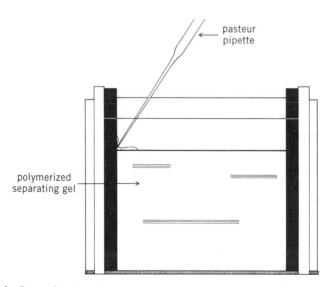

Figure 5.4. Introducing the stacking gel solution into the gel sandwich.

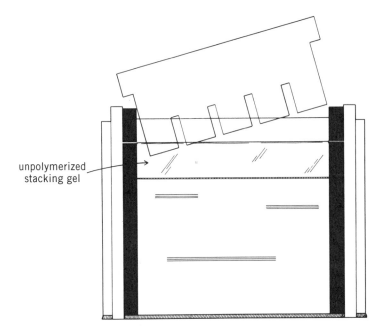

unpolymerized
stacking gel

Figure 5.5. Inserting the sample-well comb into the stacking gel.

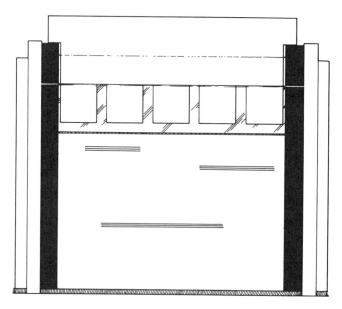

Figure 5.6. Stacking gel prior to polymerization.

Figure 5.7. Mini-Protean II electrophoresis chamber.

- **Comments**

    1. Reagent solutions are stored at 4°C and generally do not need to be warmed to room temperature before mixing and pouring gels.

    2. Possible reasons that a gel does not polymerize:

        a. Not enough ammonium persulfate or TEMED. Increase volumes of catalysts.
        b. Poor quality reagents. Use electrophoresis grade reagents.
        c. Ammonium persulfate or TEMED are inactive. Prepare or purchase fresh stock.
        d. Temperature is too low. Cast gel at room temperature.

        For a more detailed troubleshooting guide, consult Bio-Rad Bulletin 1156.

    3. Polymerization rate can be most easily adjusted by altering the amounts of polymerization catalysts used.

    4. Degassing the acrylamide solution will lead to more rapid polymerization. We have found this step to be cumbersome and unnecessary in most applications.

    5. Gel cracking during polymerization (especially high percentage gels) is likely due to excess heat generation. Use cooled reagents (Hames, 1981).

    6. If electrophoresis is to be carried out at 4°C, lithium dodecyl sulfate (LiDS) should be substituted for SDS. LiDS does not precipitate at low temperatures.

    7. Sharpness of protein bands may vary according to the brand and grade of SDS used. See Rothe and Maurer (1986, pp. 106-108) for an analysis and discussion of SDS purity from different commercial sources.

    8. For a polyacrylamide gel system which separates proteins from 50 kd to 500 kd, see Perret et al. (1983).

D. Preparing and Loading Samples

- **Capacity per Well** (Mini-Gel System)

Gel Thickness	1 Well	5 Wells	10 Wells	15 Wells
0.5 mm	0.7 ml	45 µl	16 µl	9 µl
0.75 mm	1.0 ml	68 µl	24 µl	14 µl
1.0 mm	1.4 ml	90 µl	32 µl	18 µl
1.5 mm	2.1 ml	135 µl	48 µl	27 µl

- **Steps**

    1. **Combine protein sample and 5x Sample Buffer** (i.e. 20 µl + 5 µl) in an Eppendorf tube.

    2. **Heat at 100°C for 2 - 10 minutes.**

    3. **Spin down protein solution for 1 second** in microfuge, longer if large quantities of debris are present.

    4. **Introduce sample solution into well** using a Hamilton syringe (Fig. 5.8) or disposable gel loading tip. Layer protein solution on bottom of well and raise syringe tip as dye level rises. Be careful to avoid introducing air bubbles as this may allow some of sample to be carried to adjacent well.

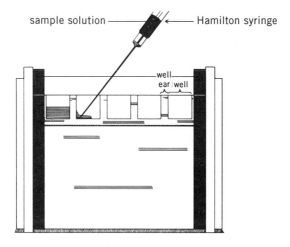

Figure 5.8. Introducing protein solution into sample well.

5

Rinse syringe thoroughly with electrode buffer or water before loading different samples.

Include molecular weight standards in one or both outside wells (see Section I. I. of this chapter).

Gels can also be loaded before inserting into the electrophoresis tank, but subsequent filling of the tank with electrophoresis buffer may cause current turbulence, resulting in cross-contamination of wells.

Although the stacking effect should reduce variations due to differences in loading volumes, it is recommended that equal sample loading volumes be used for analytical work.

To facilitate visualizing the gel wells, add bromophenol blue to a final concentration of 1 mg/ml in the stacking gel (Smith et al., 1988).

• **Comments**

1. Minimum protein loading per well (single protein band): 0.1 µg (Coomassie staining) to 2 ng (silver staining, Giulian et al., 1983).

2. Maximum protein loading per well (mixture of proteins): 20 - 40 µg.

3. Precipitation of protein in sample buffer may be due to denatured protein, too little SDS, too little reducing agent, overly acidic conditions, or presence of potassium which precipitates SDS (Hames, 1981).

4. During sample preparation, if the loading mix turns yellow, the solution has become too acidic: add NaOH until the solution turns blue; otherwise, the protein sample may migrate anomalously.

5. To avoid edge effects, add 1x sample buffer to unused wells.

6. If the sample does not sink to the bottom of the well, either there is insufficient glycerol in the sample buffer or the comb did not fit snugly, leaving polymerized acrylamide deposits in the well which block loading (Hames, 1981).

7. If boiled samples are not centrifuged prior to loading, streaking of protein bands may result. Streaking may be due to protein precipitation after accumulation of protein in the stacking gel. The aggregated proteins are thought to dissolve slowly during the course of electrophoresis. Diluting the protein sample may help (Hames, 1981).

8. Boiled samples in sample buffer are usually stable for weeks if stored frozen (-20°C), although repeated freeze-thawing may lead to protein degradation.

9. If boiled samples are stored in a freezer, the samples should be warmed prior to loading to resolubilize SDS which precipitates at low temperatures (Hames, 1981).

10. When loading wells, do not to overfill. Contaminating the adjacent well can create troublesome artifacts.

E. Running a Gel

- **Steps**

    1. **Attach electrode plugs** to proper electrodes. Current should flow towards the anode.

    2. **Turn on power supply** to 200 V (constant voltage; current will be about 100 mA at start, 60 mA at end of electrophoresis for two 0.75 mm gels; 110 mA at start, 80 mA at end for two 1.5 mm gels).

    3. The dye front should migrate to 1 - 5 mm from the bottom of the gel in 30 - 40 min for two 0.75 mm gels (40 - 50 min for 1.5 mm gels).

    Gels will become quite warm during electrophoresis although this does not affect separation. However, temperature changes will affect the migration times of proteins on a gel. Lower voltage differences result in less heat generation, slower migration times, and possibly slightly decreased resolution. Temperature also affects the migration distances of proteins relative to the bromophenol blue dye front, so it may be necessary to carry out comparative experiments under identical running voltages.

    After electrophoresis, gels may stand for a few hours before staining without harm except for gels with low percentage acrylamide in which protein will start to diffuse.

    The high electrical current used in gel electrophoresis is very dangerous. **Never disconnect electrodes before first turning off the power source**. If using an electrophoresis apparatus which is not completely shielded from the environment, always leave a clearly visible sign warning that electrophoresis is in progress.

    4. **Turn off power supply**.

    5. **Remove electrode plugs** from electrodes.

    6. **Remove gel plates** from electrode assembly.

    7. Carefully remove a spacer, and, inserting the spacer in one corner between the plates, gently **pry apart the gel plates**. The gel will stick to one of the plates.

- **Comments**

1. Only constant voltage conditions give constant protein mobility during electrophoresis (Hames, 1981).

2. Uneven band migration is often due to uneven electrical current flow. Make sure that electrophoresis buffer in both upper and lower chambers is making good contact with the gel. Other possible reasons include overloading of protein in wells, high salt content of the sample or effects at wells on the end of the gel. "Smiling," or reduced mobility of samples at the edge of the gel, may be due to uneven heat dissipation such that the center of the gel becomes hotter than the sides; decreasing the power setting may help.

F. Staining a Gel with Coomassie Blue

This method (Fig. 5.9) of staining is simple, fast, and can detect as little as 0.1 µg of protein in a single band. Generally a choice is made between using Coomassie Blue or Silver Stain (see Section G) depending on the sensitivity desired.

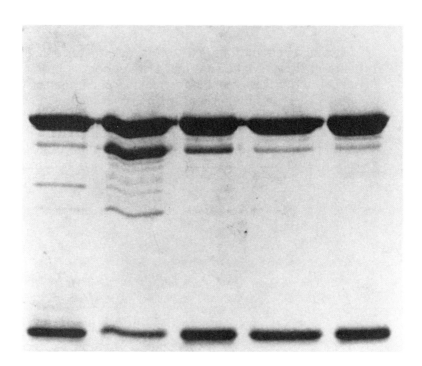

Figure 5.9. Gel stained with Coomassie Blue. The example presented is of actin (upper band) and profilin (lower band) at different stages of purification from calf spleen. The experiment was performed with a 0.75 mm gel in 13% acrylamide subjected to 200 V for 40 min. Protein concentrations were chosen to give 10 µg per well. Reprinted with permission from Rozycki et al. (1991).

- **Reagents**

    1. Coomassie Blue R-250
    2. Methanol - CH_3OH
    3. Glacial acetic acid - CH_3COOH

- **Stock Solutions**

    1. <u>Coomassie Gel Stain</u>, 1 liter
        1.0 g Coomassie Blue R-250
        450 ml methanol
        450 ml H_2O
        100 ml glacial acetic acid

    2. <u>Coomassie Gel Destain</u>, 1 liter
        100 ml methanol
        100 ml glacial acetic acid
        800 ml H_2O

- **Staining Procedure**

    1. Wearing gloves to prevent transfer of fingerprints to the gel, **pick up the gel and transfer it to a small container** (taking care not to tear the gel) containing a small amount of Coomassie Stain (20 ml is sufficient), or gently agitate the glass plate in stain solution until gel separates from plate.

    2. **Agitate for 5 - 10 minutes** for 0.75 mm, 10 - 20 min for 1.5 mm gel on slow rotary or rocking shaker. Cover container with lid or plastic wrap during staining and destaining.

    3. **Pour out stain** (can be reused several times, but it is fairly inexpensive so we generally discard it) and rinse the gel with a few changes of water. Use gloves to avoid staining hands.

    4. **Add Coomassie Destain** (about 50 ml). Strong bands are visible immediately on a light box, and the gel is largely destained within an hour. Used destain can be washed down the sink with ample water.

    5. To destain completely, change destain solution and agitate overnight. 1 - 2 cm of yarn or a piece of styrofoam can be added to absorb Coomassie stain which diffuses from the gel.

5

- **Comments**

  1. The staining and destaining times given are minimum incubation times, but staining overnight will only require longer destaining. If staining appears to be incomplete after destaining, gel can be restained.

  2. For more fragile gels (polyacrylamide concentrations less than 6%), transfer gel to staining container by placing a sheet of Whatman paper over the gel while still on the glass plate, then lift paper off with gel clinging to it and wash gel from paper in the staining container.

  3. High concentrations of SDS may interfere with Coomassie staining (Hames, 1981).

  4. Uneven staining is most likely due to incomplete penetration of the dye either because not enough dye was added or because agitation was insufficient.

  5. Nonspecific Coomassie staining may be due to deposits of undissolved dye. Filter the dye solution.

  6. Dyes tend to be attracted to positively charged groups (Lys, Arg), thus basic proteins tend to stain more strongly and some acid proteins have escaped detection (Scopes, 1982).

  7. If Coomassie staining is not sensitive enough, the gel can be rinsed and then silver stained. See silver staining comments below.

  8. A single container may be used for staining and destaining. Residual Coomassie stain can be removed from the container by wiping with a paper towel or tissue soaked in laboratory cleanser or an organic solvent such as methanol or ethanol.

G. Silver Staining a Gel

This method of staining can detect as little as 2 ng of protein in a single band (Fig. 5.10).

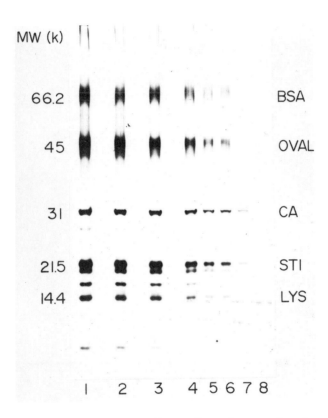

Figure 5.10. Gel stained with silver. Protein samples from the Bio-Rad low molecular weight kit were examined at dilutions of a stock solution with the proteins at the following concentrations (µg/ml): bovine serum albumin (1.2), ovalbumin (1.7), carbonic anhydrase (1.9), soybean trypsin inhibitor (1.8), and lysozyme (1.4). Lanes 1 - 8 were loaded with 2.0µl of the stock solution at the following dilutions: 1 (1:40), 2 (1:53), 3 (1:80), 4 (1:160), 5 (1:320), 6 (1:400), 7 (1:1600), and 8 (1:4000), resulting in samples in the range of 0.6 to 96 ng. Although the sensitivity varies somewhat among different proteins, generally 2 ng could be readily detected. Results from Giulian et al. (1983) kindly provided by G.G. Giulian.

- **Reagents**

    1. Silver nitrate ($AgNO_3$) (Note: Different silver nitrate batches may have different sensitivities.)
    2. Sodium hydroxide (NaOH)
    3. 14.8 M (30%) ammonium hydroxide (NH_4OH)
    4. Citric acid
    5. 38% Formaldehyde
    6. Methanol (Reagent Grade)
    7. Acetic acid
    8. Kodak rapid fix
    9. Kodak hypo clearing agent

- **Working Solutions**

    <u>Make In Advance:</u>

    1. 0.36% NaOH
    2. 1% citric acid (can be stored for several weeks)
    3. 50% methanol/10% acetic acid
    4. 1% acetic acid

    <u>Make Fresh:</u>

    5. **Solution A:** 0.8 g silver nitrate in 4 ml distilled H_2O

    6. **Solution B:**
        a. 21 ml 0.36% NaOH
        b. 1.4 ml 14.8 M (30%) ammonium hydroxide

    7. **Solution C:**
        Add Solution A to Solution B dropwise with constant vigorous stirring, allowing brown precipitate to clear. Add water to 100 ml. Use within 15 minutes.

    8. **Solution D:**
        Mix 0.5 ml 1% citric acid with 50 μl 38% formaldehyde, add water to 100 ml. Solution must be fresh.

- **Staining**

  1. Wearing gloves, pick up the gel and transfer it to a small container. **Soak gel in 50% methanol/10% acetic acid for at least 1 hour** with 2 - 3 changes of methanol/acetic acid.

  2. **Rinse 30 minutes with water**, with at least 3 changes.

  3. **Prepare Solutions A, B, then C.**

  4. **Remove gel to a clean container and stain in Solution C for 15 minutes** with gentle, constant agitation.

  5. **Rinse gel twice in deionized water, then soak 2 minutes** with gentle agitation.

  6. **Prepare Solution D.**

  7. **Remove gel to a clean container and develop by washing gel in Solution D.** Bands should appear in less than 10 minutes or else change Solution D. If a pale yellow background appears, reaction should be stopped.

  8. **Stop development by rinsing in 1% acetic acid.**

  9. **Wash gel in water for at least 1 hour** with at least three changes of water.

  10. If protein deposits are too dark, destain gel with Kodak Rapid Fix or Kodak Unifix. Stop destain with Kodak hypo clearing agent such as Orbit. Then wash in 50% methanol/10% acetic acid.

  11. Store gel in water or dry gel (see Section H, Drying a Gel).

- Notes on Staining Steps

Step 1:
- High quality methanol leads to poor staining; therefore, use reagent grade methanol.
- Initial methanol soak can be done for weeks.
- A more rapid method involves soaking the gel for 30 minutes in 70% PEG 2000 and proceeding directly to step 3 (Ohsawa and Ebata, 1983).
- Agarose-containing gels are more efficiently fixed in 20% TCA.

Step 4: **Solution C becomes highly explosive when dried.** Collect in a bottle and add an equal volume of 1 M HCl to precipitate $AgCl_2$ (Gooderham, 1984). Historically, silver chloride has been washed down the drain with ample cold water. However, increasing numbers of municipalities are regulating the introduction of heavy metal contaminants into the environment. It is best to check with your departmental waste management supervisor about the best means of disposal.

Step 7: 5 - 10% methanol can be used to slow down development.

Step 10: Destaining in Rapid Fix can also be slowed with 10% methanol.

- **Comments**

    1. Silver staining occurs mainly at gel surfaces. Therefore, to enhance surface-to-volume ratio for protein bands, use a thin gel (0.5 - 0.75 mm).

    2. If a gel is understained with Coomassie Blue, simply rinse very well with methanol and continue with step 2 of the silver stain protocol. Acetic acid interferes with silver staining, so be sure that it is completely washed out of the gel. Conversely, if silver staining overstains the gel, it can be destained with Rapid Fix and restained with Coomassie Blue.

    3. Different proteins have a nonlinear response to silver staining and basic proteins stain especially poorly. Thus, **avoid trying to estimate ratios of different proteins on silver stained gels**.

    4. Slow agitation (40 - 60 rpm) is important during incubations.

    5. Surface artifacts can be caused by pressure, fingerprints, and surface drying. Gloves should be worn at all times when handling the gel. Handling of the gel may be limited by use of a water pump to remove solutions after incubations.

    6. A uniform dark background may be due to impurities in the water (Bio-Rad Bulletin 1089). Deionized water with conductivity <1 μmho is required in all solutions.

    7. If a gray or brown precipitate appears as dust, smudges, or swirling on the gel surface, this may be due to insufficient washing during steps 2, 5, or 9, or to low temperatures (Bio-Rad Bulletin 1089).

    8. High silver staining background may be due to impurities in the acrylamide (Bio-Rad Bulletin 1156).

    9. Interfering substances such as glycerol, urea, glycine, Triton X-100, and ampholytes should be removed in the initial methanol wash (Wray et al., 1981).

10. Some variability in silver staining may be due to temperature fluctuations if the protocol is carried out at room temperature. Using a constant temperature water bath may resolve such a problem (Allen et al., 1984).

11. Silver staining may not work when the protein sample contains nucleic acids or metals. Changes in fixing and washing the gel prior to staining should help (Wedrychowski et al., 1986).

12. SDS gels in which 2-mercaptoethanol is used may develop 2 horizontal lines at 60 kd and 67 kd. These can be eliminated by using less 2-mercaptoethanol (Marshall and Williams, 1984).

13. Glutaraldehyde pretreatment has been shown to enhance staining for various proteins by up to 40-fold (Dion and Pomenti, 1983). To cross-link proteins with glutaraldehyde, wash for 30 minutes with 10% glutaraldehyde in the hood after step 1, followed by a 2 hour continuous water wash (Giulian et al., 1983). Gels can be stored for several weeks in distilled water following this step. 25% glutaraldehyde stock should be stored in the refrigerator.

14. An alternative silver staining method has been described by Merril et al. (1984).

15. Glassware must be extremely clean. The following treatments of glassware have been used successfully: acid washing, NEN Count-Off, and Malinckrodt Chem-Solv.

16. Photographs should be taken as soon as possible because of color changes and increased background with time (Gooderham, 1984).

17. ^{125}I, ^{32}P, ^{35}S, ^{14}C autoradiography after silver staining is fine, but not ^{3}H fluorography because the stain will absorb most of the emissions from the isotope (Gooderham, 1984). Destaining the gel prior to fluorography will eliminate quenching.

H. Drying a Gel

- **Materials**
  - Whatman 3MM paper
  - Acetate sheet (the kind used for transparencies) or standard plastic kitchen wrap

- **Protocol**

  1. **Place gel upside down** on a clean surface (glass plate or laboratory bench).

  2. Using water as a lubricant, adjust well ears to be upright.

  3. **Cover gel with a 10 x 12 cm piece of Whatman 3MM paper** and lift the 3MM paper with the gel off of the glass plate or benchtop. The gel should stick to the paper.

  4. **Cover the front of the gel with an acetate sheet** or plastic wrap, taking care not to trap air bubbles, which can lead to gel cracking. It is useful to roll out air bubbles, using a test tube as a rolling pin.

  5. **Place Whatman paper on gel dryer, turn on heat** (Smith, 1984, recommends 60°C) **and suction, and cover with sealing gasket**.

- **Comments**

  1. We find that a 0.75 mm gel is dry in about 1 hour and a 1.5 mm gel dries in about 2 hours, but this depends on the pump or aspirator you are using.

  2. Use paper > 1 mm thick to prevent curling (Smith, 1984).

  3. Gels can also be dried at room temperature and atmospheric pressure behind sheets of cellophane (Giulian et al., 1983 and Smith, 1984). Gels dried behind transparent sheets are useful for densitometry.

  4. Cracking of a gel during drying, especially common for high percentage acrylamide gels, may be due to release of the vacuum before the gel is completely dry (Hames, 1981). Soaking the gel overnight in 5% glycerol is often recommended to decrease the risk of cracking during drying. For other strategies to avoid gel cracking, see Smith (1984).

I. General Discussion

- Examples of abnormal migration (Weber et al., 1971):

  1. Incomplete SDS binding leads to reduced electrophoretic mobilities and occurs in the following cases:
     a. proteins which are not completely reduced;
     b. chemically cross-linked proteins;
     c. glycoproteins, and succinylated, maleylated, or acidic proteins.

  2. Proteins whose intrinsic net charge makes a significant contribution to the total net charge after SDS binding, such as histones, will have deviant relative mobilities.

  3. Proteins of molecular weight below 15 kd begin to behave unusually in SDS-PAGE, due to a charge to mass ratio that is different from larger proteins and also due to the changed properties of small particles migrating in a gel. SDS-urea gels may partially overcome these problems (see section III. below). An SDS-urea PAGE protocol for separating proteins in the 1000 to 10,000 kd range is given in Burr and Burr (1983).

- If doublets are observed where a single species is expected, it is possible that a portion of the protein sample may not be completely reduced. Increase the 2-mercaptoethanol concentration in the sample buffer. It is also reasonable to suspect proteolysis.

- Very hydrophobic proteins (for example *E. coli* lactose permease) may bind excess SDS and display high electrophoretic mobility. Higher acrylamide concentrations provide a more accurate molecular weight (See and Jackowski, 1989). Detergent-extracted proteins may require the presence of the detergent to remain soluble or active. Electrophoresis in the presence of the detergent may provide the most satisfactory solution, although detergents with a large micelle size such as Triton X-100 (see Chapter 1, Table 2) may penetrate poorly into the polyacrylamide gel (Rothe and Maurer, 1986, pp. 112-117). Also, if nonionic detergents are used, the proteins being separated will no longer have similar charge-to-mass ratios, and their electrophoretic mobilities will not necessarily be directly proportional to molecular weight.

- High concentrations of cations in the sample may cause precipitation of SDS and should be removed prior to addition of sample buffer.

- Some proteins need to be fixed rapidly or else they diffuse out of the gel (Wilson, 1983). Also, when using ultrathin gels (< 0.5 mm), fixing prior to staining is necessary. 20% TCA will also act as a fixative (Allen et al., 1984).

- Ammonium persulfate may interfere with enzyme activity. Pre-electrophoresis or substituting riboflavin may help. See Blackshear (1984).

- SDS often inhibits enzyme activity, but various strategies exist for recovering active enzymes from SDS-polyacrylamide gels. See Section IV. D. For an enzyme assay following SDS-PAGE, see Spanos and Huebscher (1983).

- For cleaning electrophoresis apparatus and plates, mild dishwashing detergent is generally adequate. Chromic acid washing followed by water and ethanol rinses is recommended by various investigators, but we obtain satisfactory results using only mild detergents, especially when staining with Coomassie Blue.

- Sealing gel plates: The Bio-Rad Mini-Protean II apparatus requires no sealing of the gel plates. If using a gel plate system which requires sealing, we recommend coating the mounted gel plates with a heated 1.5% agarose solution.

- Low percentage acrylamide gels (<3%) may be reinforced with 0.5% agarose (Hames, 1981).

- For gels with acrylamide concentrations above 12%, a higher acrylamide:bis ratio is recommended. See Blackshear (1984).

- %T and %C are terms which are often used to describe the acrylamide composition of the gel. %T refers to the total acrylamide content (w/v) while %C is the ratio of crosslinking reagent (i.e. bis-acrylamide) to acrylamide monomer (w/w). The following formulas apply:

$$\%T = \frac{\text{Acrylamide (g) + Bis (g)}}{\text{Volume (ml)}} \times 100\%$$

$$\%C = \frac{\text{Bis (g)}}{\text{Acrylamide (g) + Bis (g)}} \times 100\%$$

(from Pharmacia Publication)

- Minislab gels have become the system of choice for most electrophoretical applications due to their increased resolution and shorter running and staining times. Larger gels may, however, be useful for certain purposes such as providing greater physical separation between protein bands for autoradiography (Schleif and Wensink, 1981). On the other end of the size spectrum, the Pharmacia Phastsystem represents a milestone in reduction of size and time needed for gel electrophoresis.

- Protein stoichiometries can be determined from Coomassie stained gels by scanning densitometry and integration (see section VI. below). Various caveats apply; see Weber et al. (1971).

- A simple test for enzymatic proteolysis is to add the sample buffer to two identical protein solutions and boil one but not the other protein sample (Weber et al., 1971).

- For removal of nucleic acids, see Sinclair and Rickwood (1981) or Schleif and Wensink (1981).

- Preparation of size standards: 2 μg of a purified protein standard per lane is adequate. If preparing a mixture of standard proteins, scale up accordingly (i.e., for a mixture of 5 standard proteins, 10 μg of the size standard solution should be loaded per well). Making aliquots for each gel will help in preventing protein degradation due to repeated freeze-thawing. Prestained molecular weight markers are extremely useful, but be aware that dye binding increases the molecular weight of the markers.

J. Safety Notes

- **Acrylamide:** Extremely toxic, causing central nervous system paralysis. Also a probable carcinogen and teratogen. Can be absorbed through unbroken skin. If skin comes in contact with acrylamide powder or solution, wash with soap and much water. Unpolymerized acrylamide should be polymerized with excess catalyst and disposed of with solid waste (Merck Index and BioRad Bulletin 1156).

- **Ammonium Persulfate:** Dispose of by diluting with water (BioRad Bulletin 1156).

- **TEMED:** Store refrigerated in a dark bottle (Hames, 1981). Reported significant loss of activity after 10 - 12 months (BioRad Bulletin 1156), although we have obtained good results with stocks that are much older.

- **Silver Nitrate:** Poisonous and a skin irritant (Merck Index). Mixed silver/ammonium hydroxide solution (Solution C) dries down to form highly explosive deposits, so Ag^{2+} should be precipitated with HCl immediately after use.

- **Formaldehyde:** Vapors are very irritating. Store well-closed in a moderately warm environment (Merck Index).

II. Gradient Gels

A. Introduction

The use of a gradient of acrylamide for gel electrophoresis has two advantages over the use of a linear gel. An increased protein sieving effect at higher acrylamide concentrations leads to the formation of sharper bands at lower molecular weights. The gradient also permits the separation of a larger range of protein molecular weights in a single gel (from 15 kd to 200 kd on a 5 - 20% gel or 13 kd to 950 kd on a 3 to 30% gel). The percentages of acrylamide used in the gradient may be adjusted according to need (Hames, 1981; Walker, 1981). Precast gradient gels are commercially available.

B. Equipment

- Gradient Maker (Hoefer, Pharmacia-LKB)
- Peristaltic Pump (Baxter, Fisher, Pharmacia-LKB)
- Tubing (Baxter, Fisher)

C. Preparing Gels

- **Working Solutions**

1. Solution A - 30% acrylamide, 0.8% bis-acrylamide (Section I.C.)
2. Solution B - 1.5 M Tris-HCl (pH 8.8), 0.4% SDS (Section I.C.)
3. Solution C - 0.5 M Tris-HCl (pH 6.8), 0.4% SDS (Section I.C.)

- **Amounts of Solutions for a 5 - 20% Separating Gel**

5%	20%	
1.67 ml	6.67 ml	Solution A
2.5 ml	2.5 ml	Solution B
5.8 ml	--	H_2O
--	1.5 g	Sucrose (adds 0.8 ml to volume)
50 μl	50 μl	10% Ammonium Persulfate
5 μl	5 μl	TEMED

• Instructions for Forming Separating Gel

The only significant difference in pouring a gradient acrylamide gel as opposed to a linear gel is the use of a gradient maker. A gradient maker (Fig. 5.11) mixes the high and low concentration acrylamide solutions just prior to pouring the separating gel mixture into the gel sandwich. In the absence of a peristaltic pump, the gel solution may be introduced into the gel sandwich by gravity. Sucrose is also added to the high concentration acrylamide solution to stabilize gradient formation.

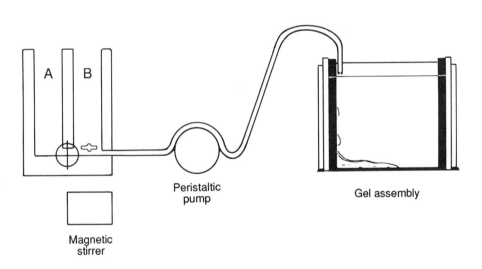

Figure 5.11. Apparatus for pouring gradient gels.

1. Prepare the gel sandwich and set up the gradient maker as shown in Fig. 5.11. The magnetic stirrer should be isolated from the gradient maker so as not to heat the mixing chambers (i.e., with a sheet of styrofoam). Calibrate the peristaltic pump to a flow rate of approximately 3 ml/min.

2. Prepare the separating gel solutions without adding TEMED.

5

3. Add TEMED to the separating gel solutions and mix gently. Immediately transfer the appropriate volume of each solution into the mixing chambers (i.e., for a 10 ml gel, add 5 ml into each mixing chamber). The high concentration (20%) acrylamide solution should be added to chamber B. **Note: It is very important to work rapidly from this point onward because polymerization will be under way**.

4. Turn on the magnetic stirrer and open the connection between the two chambers.

5. Turn on the peristaltic pump and allow gradient gel to form.

6. When the separating gel has been poured, gently layer about 1 to 5 mm of water on top of the gel.

7. Wash the gradient maker and tubing with water to prevent the acrylamide solution from polymerizing inside.

8. Instructions for the preparation of the stacking gel can be found in section I.C.

- **Comments**

  1. Air bubbles lodged in the tubing or between the gradient mixing chambers can cause the gradient to form unevenly.

  2. Gel solutions may be chilled to allow more time for pouring the gradient before polymerization occurs.

  3. Riboflavin may be substituted for ammonium persulfate to allow more time for pouring the gradient before polymerization occurs. A final riboflavin concentration of 0.0005% (from a 0.004% stock solution) may be used, and polymerization is initiated by exposure to daylight or a white or blue fluorescent lamp.

Consult: Hames (1981) or Walker (1984).

III. SDS-Urea Gels

A. Introduction

The mobilities of small proteins in SDS-PAGE may no longer be proportional to their molecular weight when the protein charge properties become significant relative to the mass. SDS-urea PAGE is often used in these cases (Swank and Munkres, 1971; Schleif and Wensink, 1981; Burr and Burr, 1983). In addition, SDS-urea PAGE may be useful with proteins such as immunoprecipitates and membrane proteins, which are not soluble at low ionic strengths.

Aside from the gel composition described below, all procedures are as described for SDS-PAGE (Section I.).

B. Preparing Gels

- **Working Solutions**
    1. Solution A - 30% acrylamide, 0.8% bis-acrylamide (Section I.C.)
    2. Solution B - 1.5 M Tris-HCl (pH 8.8), 0.4% SDS (Section I.C.)
    3. Solution C - 0.5 M Tris-HCl (pH 6.8), 0.4% SDS (Section I.C.)

- **Amounts of Solutions**

For X% Separating Gel (10 ml)		X% Stacking Gel (4 ml)
Solution A	$x/3$ ml	$x/7.5$ ml
Solution B	2.5 ml	0 ml
Solution C	0 ml	1 ml
Urea	4.8 g =3.6 ml	1.9 g = 1.4 ml
H_2O	3.9 - $x/3$ ml	1.6 - $x/7.5$ ml
10% Am. Persulfate	50 µl	30 µl
TEMED	5 µl	5 µl

Note: These formulas for separating and stacking gels are valid only up to an acrylamide concentration of 12%. For higher acrylamide concentrations, urea should be included in Solutions A and B. See Schleif and Wensink (1981).

- **Example**

8% Separating Gel (10 ml)   **5% Stacking Gel** (4 ml)

	8% Separating Gel	5% Stacking Gel
Solution A	2.7 ml	0.67 ml
Solution B	2.5 ml	0 ml
Solution C	0 ml	1.0 ml
Urea	4.8 g	1.9 g
H_2O	1.2 ml	0.93 ml
10% Am. Persulfate	50 µl	30 µl
TEMED	5 µl	5 µl

Electrophoresis conditions and buffer solutions as well as staining and destaining solutions are the same as described for SDS-PAGE (Section I.). 5x Sample buffer should be made to contain 8 M urea.

IV. Other Methods

A. Detection of Radiolabeled Samples

The use of radioactively labeled proteins for gel electrophoresis has extended the limits of protein detection. Radioactivity can be measured by autoradiography (exposure of the gel to X-ray film) or by cutting gel slices and counting them in a scintillation counter.

Gel slicing is a more quantitative assay, although it suffers from poor resolution. Gels may be sliced with a razor blade, and commercial gel slicers are also available (Bio-Rad, Hoefer). For good liquid scintillation counting efficiency, proteins should first be eluted from the gel slice. A simple procedure for dissolving gel slices with H_2O_2 is described by Hames (1981, pp. 56-58).

Autoradiography may involve direct exposure of the X-ray film to the gel (sealed in plastic wrap), or it may be enhanced more than tenfold by the use of an intensifying screen behind the film (called indirect autoradiography). The sensitivity of detection for ^{14}C, ^{32}P, and ^{35}S is enhanced approximately 15-fold by the use of fluorographic methods. Among the more popular of these is gel impregnation with a fluor such as EN[3]HANCE (New England Nuclear). Fluorography is necessary for detection of [3]H. Methods for autoradiography and fluorography are described in the references listed below.

Consult: Bonner (1984), Hames (1981), Harlow and Lane (1988), Roberts (1985), or Waterborg and Matthews (1984).

B. Molecular Weight Determination

SDS-PAGE is frequently used to determine the molecular weight of a protein since protein migration is generally proportional to the mass of the protein. A standard curve is generated with proteins of known molecular weight, and the molecular weight of the protein of interest can be extrapolated from this curve.

- **Protocol**

    1. Following gel electrophoresis and staining, measure the distance of migration of the proteins as well as that of the tracking dye (bromophenol blue). Distance of migration is measured from the beginning of the separating gel to the leading edge of a protein band.

    2. Calculate R_f values

    $$R_f = \frac{\text{distance of protein migration}}{\text{distance of tracking dye migration}}$$

    3. Plot the $\log_{10}$ of the known protein molecular weights as a function of their R_f. The area in the middle of the gel should yield a straight line (Fig. 5.12).

    4. Read molecular weight of the unknown protein from the graph based on its R_f value.

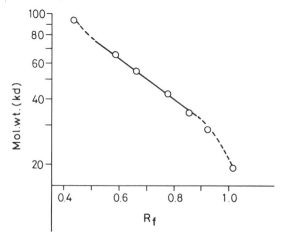

Figure 5.12. Semilogarithmic graph of molecular weight versus relative mobility. Modified from See and Jackowski (1989).

- **Comments**

1. Calculations of R_f values should be made from proteins separated on the same slab gel. This eliminates variability due to acrylamide concentration and electrophoretic conditions.

2. Molecular weight determinations based solely on SDS-PAGE may be misleading, since some proteins migrate anomalously.

3. For a careful molecular weight determination, it is advisable to run gels of at least two different acrylamide concentrations.

4. A new standard curve must be generated for each gel.

5. Protein size standards are available commercially. A table of selected protein molecular weights and isoelectric points is provided in Appendix 2.

6. Be careful to make appropriate molecular weight corrections for prestained molecular weight markers. Dye binding to standard proteins increases their molecular weight. Consult the specification sheet accompanying the prestained markers.

7. Gels may be dried prior to measurement of R_f values. Keep in mind, however, that drying tends to broaden protein bands, especially for thicker gels, and may make it more difficult to find the exact position of the leading edge.

Consult: Hames (1981) or See and Jackowski (1989).

C. Protein Quantitation (Densitometry)

Quantitation of amounts of individual proteins in a mixture is often accomplished by scanning stained protein bands on a polyacrylamide gel with a spectrophotometer, and then integrating the peaks. Peak integration may be achieved electronically with some densitometers or manually by cutting the peaks from the chart paper and weighing. The absorbance of a protein band is proportional to the amount of protein only over a limited range of protein concentration. Proteins bind dyes with varying affinity, so comparison of peaks from different proteins should be based on standard curves for absolute quantitation when possible. Smith (1984) provides suggestions for the analysis of densitometric scans. Commercial gel scanning devices are available from Bio-Rad, Hoefer, Joyce-Loebl, and Pharmacia-LKB.

Protein quantitation may also involve elution of dye from stained gel bands. Ball (1986) and Wong et al. (1985) describe simple methods for removing Coomassie Blue R-250 from polyacrylamide gel slices. It is important to note that absolute quantitation requires the generation of a standard curve with the protein of interest due to differences in dye binding by different proteins. The procedure of Ball (1986) is presented below:

- **Protocol**

    1. Cut out protein band from the gel with a razor blade.

    2. Place gel slice in a glass test tube, add 1 ml of 3% SDS in 50% isopropanol, and cover with Parafilm.

    3. Incubate in a 37°C water bath for 24 hours without agitation.

    4. Remove liquid and determine absorbance at 595 nm.

Consult: Ball (1986), Smith (1984), or Wong et al. (1985).

D. Eluting Protein Bands Following Electrophoresis

Extraction of protein from acrylamide gels can be accomplished by passive protein diffusion or electroelution Harrington, 1990). A simple procedure for passive protein elution (Bhown and Bennett, 1984) involves cutting the gel slice into small pieces with a razor, incubating for various times (15 minutes is a good period) in an appropriate buffer, followed by centrifugation, collection of the supernatant, and repeating the extraction a second time with more buffer. A better recovery yield can be expected with electroelution. A simple electroelution procedure termed reverse electrophoresis employing a tube gel is described by Otto and Snejdarkova (1981) and is adapted for slices from slab gels by Stralfors and Belfrage (1983). Electroelution devices are also available commercially from Bio-Rad and Pharmacia-LKB.

Reconstitution of enzyme activity following SDS-PAGE has been reported both *in situ* and after protein elution from gel slices (see also Chapter 6, Section III.B.). SDS removal sometimes requires replacement by a less denaturing detergent (nonionic or zwitterionic, see Chapter 1, Section IV.). Successful renaturation may depend on the concentrations of salt, glycerol, reducing agent, cofactor, or substrate in the renaturation buffer. A good discussion of enzyme renaturation following SDS-PAGE is found in Rothe and Maurer (1986, pp. 108-112). (See also the discussion of protein renaturation from bacterial inclusion bodies in Chapter 2, Section III.)

Consult: Bhown and Bennett (1983), Otto and Snejdarkova (1981),
     Rothe and Maurer (1986), or Stralfors and Belfrage (1983).

V. Suppliers

- **Equipment**

  Bath, boiling water 100°C sand bath: Baxter

  Centrifuge, Eppendorf: Baxter, Fisher

  Gel dryer:
  - Bio-Rad Model 443 and 483
  - Hoefer Drygel
  - Pharmacia GSD-4 Gel Slab Dryer

  Minigel apparatus:
  - Bio-Rad Mini-Protean II Electrophoresis Cell
  - Hoefer Scientific
  - Pharmacia Midget System

  Power supply (capacity 200V, 400mA):
  - Bio-Rad Model 250/2.5 Power Supply
  - Pharmacia EPS 500/400 or GPS 200/400 Power Supply

  Rocking or rotary shaker: Hoefer

  Syringes: Hamilton

  Vacuum, water pumps: Baxter, Hoefer

- **Reagents**

  Precast gels: Bio-Rad, ISS, Novex, Pharmacia

  Electrophoresis Reagents: Aldrich, Bio-Rad, Calbiochem, Fluka, Merck, Pharmacia, Sigma, Whatman

  Silver Staining Kits: Bio-Rad, Pierce, Polysciences

VI. References

SDS Polyacrylamide Gel Electrophoresis

Allen, R.C., C.A. Saravis, and H.R. Maurer. 1984. Gel Electrophoresis and Isoelectric Focusing of Proteins: Selected Techniques. 255 pages. Walter de Gruyter, Berlin.

Bio-Rad Bulletin 1089. 1984. Bio-Rad Silver Stain.

Bio-Rad Bulletin 1156. 1984. Acrylamide Polymerization - A Practical Approach.

Bio-Rad Mini Protean Slab Gel Instruction Manual.

Blackshear, P.J. 1984. Meth. Enzymol. 104: 237-255. Systems for Polyacrylamide Gel Electrophoresis.

Dion, A.S. and A.A. Pomenti. 1983. Anal. Biochem. 129: 490-496. Ammoniacal Silver Staining of Proteins: Mechanism of Glutaraldehyde Enhancement.

Giulian, G.G., R.L. Moss and M. Greaser. 1983. Anal. Biochem. 129: 277-287. Improved Methodology for Analysis and Quantitation of Proteins on One-Dimensional Silver-Stained Slab Gels.

Gooderham, K. 1984. pp. 113-118 in Methods in Molecular Biology. Volume 1: Proteins. J.M. Walker, ed. 365 pages. Humana Press, Clifton, New Jersey.

Hames, B.D. pp. 1-91 in Hames, B.D and D. Rickwood, eds. 1981. Gel Electrophoresis of Proteins: A Practical Approach. 290 pages. IRL Press, Oxford and Washington, D.C.

Laemmli, U.K. 1970. Nature 227: 680-685. Cleavage of Structural Proteins during the Assembly of the Head of Bacteriophage T4.

Marshall, Thomas and Katherine M. Williams. 1984. Anal. Biochem. 139: 502-505. Artifacts Associated with 2-Mercaptoethanol upon High Resolution Two-Dimensional Electrophoresis.

Merck Index, The. 1983. 10th Edition. M. Windholz, Ed. Merck & Co., Inc. Rahway, NJ.

Merril, C.R., D. Goldman and M.L. Van Keuren. 1984. Meth. Enzymol. 104: 441-446. Gel Protein Stains: Silver Stain.

Ohsawa, K. and N. Ebata. 1983. Anal. Biochem. 135: 409-415. Silver Stain for Detecting 10-Femtogram Quantities of Protein after Polyacrylamide Gel Electrophoresis.

Perret, B.A., R. Felix, M. Furlan and E.A. Beck. 1983. Anal. Biochem. 131: 46-50. Silver Staining of High Molecular Weight Proteins on Large-Pore Polyacrylamide Gels.

Pharmacia Publication. Polyacrylamide Gel Electrophoresis: Laboratory Techniques. 72 pages.

Rothe, G.M. and W.D. Maurer. 1986. pp. 37-140 in Gel Electrophoresis of Proteins. 407 pages. M.J. Dunn, ed. IOP Publishing Limited, Bristol, England.

Rozycki, M., C.E. Schutt and U. Lindberg. 1991. Meth. Enzymol. 196: 100-118. Affinity Chromatography-Based Purification of Profilin:Actin.

Schleif, R.F. and P.C. Wensink. 1981. pp. 78-84. Practical Methods in Molecular Biology. 220 pages. New York, Springer-Verlag.

Scopes, R.K. 1982. pp. 163-171, 245-254. Protein Purification: Principles and Practice. 282 pages. Springer-Verlag, New York.

See, Y.P. and G. Jackowski. 1989. pp. 1-22 in Protein Structure: A Practical Approach. 355 pages. T.E. Creighton, ed. IRL Press, Oxford, England.

Sinclair, J. and D. Rickwood. pp. 189-218 in Hames, B.D. and D. Rickwood, eds. 1981. Gel Electrophoresis of Proteins: A Practical Approach. 290 pages. IRL Press, Oxford and Washington, D.C.

Smith, B.J. 1984. p. 141-146 in Methods in Molecular Biology. Volume 1: Proteins. J.M. Walker, ed. 365 pages. Humana Press, Clifton, New Jersey.

Smith, I., R. Cromie and K. Stainsby. 1988. Anal. Biochem. 169: 370-371. Seeing Gel Wells Well.

Spanos, A. and U. Huebscher. 1983. Meth. Enzymol. 91: 263-277. Recovery of Functional Proteins in Sodium Dodecyl Sulfate Gels.

Weber, K., J.R. Pringle and M. Osborn. 1971. Meth. Enzymol. 26: 3-27. Measurement of Molecular Weights by Electrophoresis on SDS-Acrylamide Gel.

Wedrychowski, A., R. Olinski and L.S. Hnilica. 1986. Anal. Biochem. 159: 323-328. Modified Method of Silver Staining of Proteins in Polyacrylamide Gels.

Wilson, C.M. 1983. Meth. Enzymol. 91: 236-246. Staining of Proteins on Gels: Comparisons of Dyes and Procedures.

Wray, W., T. Boulikas, V.P. Wray and R. Hancock. 1981. Anal. Biochem. 118: 197-203. Silver Staining of Proteins in Polyacrylamide Gels.

Gradient Gels

Hames, B.D. 1981. pp. 71-77 in Gel Electrophoresis of Proteins: A
   Practical Approach. Hames, B.D. and D. Rickwood, eds. 290
   pages. IRL Press, Oxford and Washington D.C.
Walker, J.M. 1984. pp. 57-62 in Methods in Molecular Biology.
   Volume 1: Proteins. J.M. Walker, ed. 365 pages. Humana Press,
   Clifton, New Jersey.

SDS-Urea Gels

Burr, F.A. and B. Burr. 1983. Meth. Enzymol. 96: 239-244. Slab Gel
   System for the Resolution of Oligopeptides below Molecular
   Weight of 10,000.
Schleif, R.F. and P.C. Wensink. 1981. pp. 84-87 in Practical Methods
   in Molecular Biology. 220 pages. Springer-Verlag, New York.
Swank, R.T. and K.D. Munkres. 1971. Anal. Biochem. 39: 462-477.
   Molecular Weight Analysis of Oligopeptides by Electrophoresis in
   Polyacrylamide Gel with Sodium Dodecyl Sulfate.

Detection of Radiolabeled Samples

Bonner, W.M. 1984. Meth. Enzymol. 104: 460-466. Fluorography
   for the Detection of Radioactivity in Gels.
Hames, B.D. 1981. pp. 49-59 in Gel Electrophoresis of Proteins: A
   Practical Approach. Hames, B.D. and D. Rickwood, eds. 290
   pages. IRL Press, Oxford and Washington, D.C.
Harlow, E. and D. Lane. 1988. pp. 647-648 in Antibodies: A
   Laboratory Manual. 726 pages. Cold Spring Harbor Laboratory,
   Cold Spring Harbor, New York.
Roberts, P.L. 1985. Anal. Biochem. 147: 521-524. Comparison of
   Fluorographic Methods for Detecting Radioactivity in
   Polyacrylamide Gels or on Nitrocellulose Filters.
Waterborg, J.H. and H.R. Matthews. 1984. pp. 147-152 in Methods
   in Molecular Biology. Volume 1: Proteins. J.M. Walker, ed. 365
   pages. Humana Press, Clifton, New Jersey.

Molecular Weight Determination

Hames, B.D. 1981. pp. 14-17 in Gel Electrophoresis of Proteins: A Practical Approach. Hames, B.D. and D. Rickwood, eds. 290 pages. IRL Press, Oxford and Washington, D.C.

See, Y.P. and G. Jackowski. 1989. pp. 10-18 in Protein Structure: A Practical Approach. T.E. Creighton, ed. 355 pages. IRL Press, Oxford.

Protein Quantitation (Densitometry)

Ball, E.H. 1986. Anal. Biochem. 155: 23-27. Quantitation of Proteins by Elution of Coomassie Brilliant Blue R from Stained Bands after Sodium Dodecyl Sulfate - Polyacrylamide Gel Electrophoresis.

Smith, B.J. 1984. pp. 119-125 in Methods in Molecular Biology. Volume 1: Proteins. J.M. Walker, ed. 365 pages. Humana Press, Clifton, New Jersey.

Wong, P., A. Barbeau and A.D. Roses. 1985. Anal. Biochem. 150: 288-293. A Method to Quantitate Coomassie Blue-Stained Proteins in Cylindrical Polyacrylamide Gels.

Eluting Protein Bands Following Electrophoresis

Bhown, A.S. and J.C. Bennett. 1983. Meth. Enzymol. 91: 450-455. High-Sensitivity Sequence Analysis of Proteins Recovered from Sodium Dodecyl Sulfate Gels.

Harrington, M.G. 1990. Meth. Enzymol. 182: 488-495. Elution of Protein from Gels.

Otto, M. and M. Snejdarkova. 1981. Anal. Biochem. 111: 111-114. A Simple and Rapid Method for the Quantitative Isolation of Proteins from Polyacrylamide Gels.

Rothe, G.M. and W.D. Maurer. 1986. pp. 37-140 in Gel Electrophoresis of Proteins. M.J. Dunn, ed. 407 pages. IOP Publishing Ltd., Bristol, England.

Stralfors, P. and P. Belfrage. 1983. Anal. Biochem. 128: 7-10. Electrophoretic Elution of Proteins from Polyacrylamide Gel Slices.

Chapter 6

Gel Electrophoresis under Nondenaturing Conditions

6

I.   Introduction

II. Discontinuous Nondenaturing Gel Electrophoresis
     A. Introduction
     B. Equipment
     C. Preparing the Gel
     D. Sample Preparation
     E. Running the Gel
     F. Staining the Gel - Coomassie Blue
     G. A Variant: Continuous Nondenaturing Gel Electrophoresis

III. Related Methods
     A. Determining Protein Molecular Weight
     B. Determining Enzyme Activity after Electrophoresis

IV. References

I. Introduction

Nondenaturing gel electrophoresis, also called native gel electrophoresis, separates proteins based on their size and charge properties. While the acrylamide pore size serves to sieve molecules of different sizes, proteins which are more highly charged at the pH of the separating gel have a greater mobility. This method is capable of separating molecules which differ by a single unit charge. In addition, the conditions for nondenaturing gel electrophoresis minimize protein denaturation, in contrast to SDS polyacrylamide gel electrophoresis described in Chapter 5.

It is important to appreciate the effects of a few variables on nondenaturing gel electrophoresis.

Gel porosity: Acrylamide concentrations in the gel may be varied from approximately 5% to 15% and acrylamide:bis-acrylamide ratios may vary from 20:1 to 50:1 to achieve different sieving effects.

Charge: Most proteins are negatively charged at pH 8.8, which is the common pH used for nondenaturing gel electrophoresis. Alternatively, electrophoresis may be carried out at slightly acidic pH, in which case the anode and cathode should be reversed. Note that in order to recover biological activity, it is necessary to work in a pH range which is not harmful to the protein of interest. For a practical discussion of pH effects on electrophoresis, see Hames (1981).

Ionic Strength: The ionic strength plays an important role in electrophoresis. If the ionic strength is too high there will be increased heat generation during electrophoresis but if the ionic strength is too low proteins may aggregate nonspecifically. Typically, ionic strengths in the 10 - 100 mM range are utilized.

Temperature: All steps should be performed at 0 - 4°C to reduce loss of protein activity by denaturation and to minimize attack by proteolysis.

II. Discontinuous Nondenaturing Gel Electrophoresis

A. Introduction

Nondenaturing gel electrophoresis is commonly run with high pH buffers (pH 8.8). At this pH, most proteins are negatively charged and migrate toward the anode. The instructions are written for the Bio-Rad Mini-Protean II apparatus, but all protocols should be easily adaptable to other systems.

6

B. Equipment

- Bio-Rad Mini-Protean II Gel Electrophoresis Cell

- Power Supply

- Hamilton syringe or disposable gel loading tip for micropipette

- Small container for staining and destaining the gel

Consult Chapter 5 for a list of suppliers.

C. Preparing the Gel

- **Reagents**

    1. Acrylamide
    2. Bis-acrylamide
    3. Tris
    4. Hydrochloric acid (HCl)
    5. Ammonium persulfate
    6. TEMED
    7. Glycine
    8. Glycerol
    9. Bromophenol blue
    10. Coomassie Blue R-250 (for Coomassie Blue staining)
    11. Methanol
    12. Glacial acetic acid

- **Working Solutions**

1. **Solution A** (Acrylamide Stock Solution), 100 ml

   30% Acrylamide, 0.8% Bis-acrylamide

   **Caution: Unpolymerized acrylamide is a skin irritant and a
   neurotoxin. Always handle with gloves (see comments
   in Chapter 5).**

   a. 30 g acrylamide
   b. 0.8 g bis-acrylamide

   Add distilled water to make 100 ml and stir until completely
   dissolved. Work under hood and keep acrylamide solution
   covered with Parafilm until acrylamide powder is completely
   dissolved.

2. **Solution B** (4x Separating Buffer), 100 ml

   1.5 M Tris-HCl (pH 8.8)

   a. 18.2 g Tris in 40 ml H_2O
   b. Add HCl to pH 8.8
   c. Add H_2O to 100 ml

3. **Solution C** (4x Stacking Buffer), 100 ml

   0.5 M Tris (pH 6.8)

   a. 6.0 g Tris-HCl in 40 ml H_2O
   b. Add HCl to pH 6.8
   c. Add H_2O to 100 ml

4. **10% Ammonium Persulfate**, 5 ml

   a. 0.5 g ammonium persulfate
   b. 5 ml H_2O

6

5. **Electrophoresis Buffer**, 1 liter

    a. 3.0 g Tris   ---25 mM
    b. 14.4 g glycine   ---192 mM
    c. H_2O to 1 liter (final pH should be 8.8)

6. **5x Sample Buffer**, 10 ml

    a. 3.1 ml 1 M Tris-HCl (pH 6.8) ---312.5 mM
    b. 5 ml glycerol   ---50%
    c. 0.5 ml 1% bromophenol blue ---0.05%
    d. 1.4 ml H_2O

6

- **Amounts of Working Solutions to Use**

1. Volumes necessary for pouring gels of different thicknesses
(for two 6 x 8 cm gels)

Gel Thickness	Separating	Stacking
0.5 mm	5.6 ml	1.4 ml
0.75 mm	8.4 ml	2.1 ml
1.0 mm	11.2 ml	2.8 ml
1.5 mm	16.8 ml	4.2 ml

Always prepare with a moderate excess.

2. Calculation for X% Separating Gel

Solution A	$^x/_3$ ml
Solution B	2.5 ml
H_2O	$(7.5-^x/_3)$ ml
10% Ammonium Persulfate	50 µl
TEMED	5 µl (10 µl if x<8%)
—————————————	
Total Volume	10 ml

- **Pouring the Separating Gel**

    Example of Separating Gel Preparation

    Two 8% Separating Gels (for a 6 cm x 8 cm x 0.75 mm gel, need to prepare 10 ml)

    4.8 ml H_2O
    2.7 ml Solution A
    2.5 ml Solution B

    ---

    50 μl 10% Ammonium Persulfate
    5 μl TEMED

    Do not prepare until following the numbered instructions below.

    The instructions given below are very similar to those for denaturing gel electrophoresis in Chapter 5. Consult Chapter 5 for greater detail.

1. **Assemble gel sandwich** according to the manufacturer's instructions.

2. **Combine Solutions A and B and water** in a small Erlenmeyer flask or a test tube. Acrylamide (in Solution A) is a neurotoxin, so plastic gloves should be worn at all times.

3. **Add ammonium persulfate and TEMED, and mix** by swirling or inverting container gently. Work rapidly at this point because polymerization will be under way.

4. **Introduce solution into gel sandwich** using a pipette (Fig. 6.1).

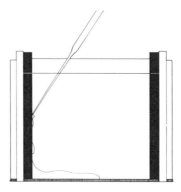

Figure 6.1. Introducing the separating gel solution into the gel sandwich.

5. When the appropriate amount of separating gel solution has been added, gently **layer about 1 - 5 mm of water** on top of the separating gel solution.

6. **Allow gel to polymerize** (30 - 60 minutes).

   When the gel has polymerized, a distinct interface will appear between the separating gel and the water, and the gel mold can be tilted to verify polymerization.

6

- **Comments on Separating Gels**

1. A low pH gel system for separation of basic proteins has been described by Reisfeld et al. (1962) with the following buffer composition (see discussion in Chapter 5, Section I.I for definitions of %T and %C terms):
   Separating gel: 0.06 N KOH, 0.376 M acetic acid, pH 4.3
      (7.7%T, 2.67%C)
   Stacking gel: 0.06 N KOH, 0.063 M acetic acid, pH 6.8
      (3.125%T, 25%C)
   Electrophoresis buffer: 0.14 M β-alanine, 0.35 M acetic acid,
      pH 4.5.

2. Remember to reverse the electrode polarity for low pH gels.

3. For the low pH gel system, methyl green should be used as the tracking dye in the sample buffer (at 0.002% final concentration).

4. Riboflavin may be used instead of ammonium persulfate for photopolymerization. Polymerization with riboflavin will leave fewer reactive compounds, but polymerization may not be as complete. A final riboflavin concentration of 0.0005% (from a 0.004% stock solution) is often utilized in the stacking gel, and polymerization is initiated by exposure to daylight or a white or blue fluorescent lamp.

5. If a reducing agent is required, 1 mM DTT is recommended.

- **Pouring the Stacking Gel**

Example of Stacking Gel Preparation

> **Two 5% Stacking** (for a 6 cm x 8 cm x 0.75 mm
>   gel, need to prepare 4 ml)
> 2.3 ml H_2O
> 0.67 ml Solution A
> 1.0 ml Solution C
> _____
>
> 30 µl 10% Ammonium Persulfate
> 5 µl TEMED
>
> **Do not prepare until following instructions in section below.**

1. **Pour off water** covering the separating gel.

2. **Combine Solutions A and C and water** in a small Erlenmeyer flask or a test tube.

3. **Add ammonium persulfate and TEMED, and mix** by gently swirling or inverting the container.

4. **Pipet stacking gel solution onto separating gel** until solution reaches top of front plate (Fig. 6.2).

5. **Carefully insert comb** into gel sandwich until bottom of teeth reach top of front plate (Figs. 6.3 and 6.4). Be sure no bubbles are trapped on ends of teeth.

6. **Allow stacking gel to polymerize** (about 30 minutes).

7. After stacking gel has polymerized, **remove comb** carefully (making sure not to tear the well ears).

8. Place gel into electrophoresis chamber. If using the Mini-Gel system, **attach both gels to electrode assembly before inserting into electrophoresis tank.**

9. **Add electrophoresis buffer to inner and outer reservoir**, making sure that both top and bottom of gel are immersed in buffer.

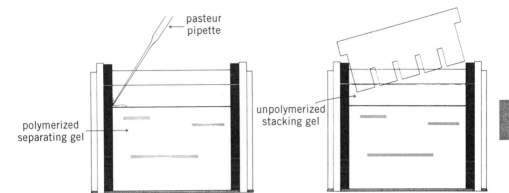

Figure 6.2. Introducing the stacking gel solution into the gel sandwich, as presented in Fig. 5.4.

Figure 6.3. Inserting the sample well comb into the stacking gel, as presented in Fig. 5.5.

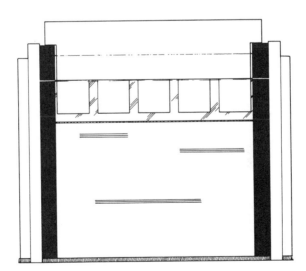

Figure 6.4. Stacking gel prior to polymerization, as presented in Fig. 5.6.

D. Sample Preparation

- **Capacity per Well** (Mini-Gel System)

Gel Thickness	1 Well	5 Wells	10 Wells	15 Wells
0.5 mm	0.7 ml	45 μl	16 μl	9 μl
0.75 mm	1.0 ml	68 μl	24 μl	14 μl
1.0 mm	1.4 ml	90 μl	32 μl	18 μl
1.5 mm	2.1 ml	135 μl	48 μl	27 μl

- **Steps**

1. **Combine protein sample and 5x Sample Buffer** (e.g., 20 μl + 5 μl) in an Eppendorf tube.

2. **Introduce sample solution into well** using a Hamilton syringe (Fig. 6.5) or disposable gel loading tip.

Rinse syringe thoroughly with electrode buffer or water before loading different samples.

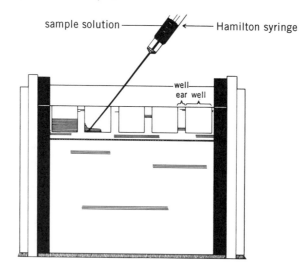

Figure 6.5. Introducing protein solution into sample well, as presented in Fig. 5.8.

- **Comments on Sample Preparation**

  1. Typically, 1 - 5 μg of protein is loaded per well, or up to 30 μg of a complex protein mixture.

  2. It is a good practice to fill unused gel lanes with blank sample buffer.

  3. Sample may be centrifuged for 15 min at 10,000 x g prior to preparation for loading in order to remove insoluble material which may interfere with electrophoresis (Hames).

  4. Samples with salt concentrations in excess of 0.1 M may cause band distortion.

6

E. Running the Gel

- **Steps**

1. **Attach electrode plugs** to proper electrodes. Current should flow towards the anode for pH 8.8 gels.

2. **Turn on power supply** to 100 - 200 V (constant current).

3. Electrophoresis should continue until the dye front migrates to within 1 - 5 mm of the bottom of the gel.

The high electrical current used in gel electrophoresis is very dangerous. **Never disconnect electrodes before first turning off the power source**. If using an electrophoresis apparatus which is not completely shielded from the environment, always leave a clearly visible sign warning that electrophoresis is in progress.

---

4. **Turn off power supply**.

5. **Remove electrode plugs** from electrodes.

6. **Remove gel plates** from electrode assembly.

7. Carefully remove a spacer, and, inserting the spacer in one corner between the plates, gently **pry apart the gel plates**. The gel will stick to one of the plates.

- **Comments**

1. Constant voltage will allow constant protein mobility during electrophoresis.

2. If the current is too high, excess heating may denature the protein, while current which is too low increases the time of electrophoresis and diffusion of bands. 100 - 200 V is the recommended range for electrophoresis.

F. Staining the Gel

We describe Coomassie Blue staining of nondenaturing gels (see Fig. 6.6). Coomassie Blue staining can detect as little as 0.1 μg of protein in a single band. Greater sensitivity is possible with the use of silver staining. For information about silver staining and more information about Coomassie Blue staining, see Chapter 5, Section I.F.

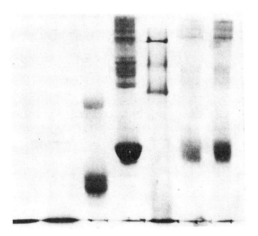

6

Figure 6.6. Nondenaturing gel stained with Coomassie Blue. The gel concentration is 6% and was run with an assortment of protein markers from the Sigma nondenatured protein molecular weight marker kit.

- **Stock Solutions** (same as for SDS-PAGE)

1. **Coomassie Gel Stain**, 1 liter
    1.0 g Coomassie Blue R-250
    450 ml methanol
    450 ml H$_2$O
    100 ml glacial acetic acid

2. **Coomassie Gel Destain**, 1 liter
    100 ml methanol
    100 ml glacial acetic acid
    800 ml H$_2$O

- **Staining Procedure**

1. Wearing gloves, **pick up the gel and transfer to a small container** containing Coomassie Stain (20 ml is sufficient).

2. **Agitate for 5 - 10 minutes** for 0.75 mm gels or 10 - 20 min for 1.5 mm gels on a slow rotary or rocking shaker.

3. **Pour out stain**.

4. **Add Coomassie Destain** (about 50 ml) and continue slow shaking.

5. To completely destain, change destain solution and agitate overnight.

Instructions for gel drying can be found in Chapter 5 (Section I.H.).

G. Variation: Continuous Nondenaturing Gel Electrophoresis

Continuous gel electrophoresis is somewhat easier to perform than discontinuous gel electrophoresis since no stacking gel is involved. Instead, the gel sandwich is filled with separating gel solution only before the comb is inserted. However, the lack of stacking gel often results in thick, poorly resolved bands. It is important in continuous gel electrophoresis that the ionic strength of the protein buffer be 5 - 10 times less concentrated than the gel buffer in order to obtain the sharpest bands. Protein samples should fill no more than two or three millimeters of the well.

Buffers for continuous gel electrophoresis may be the same as described for discontinuous gel electrophoresis, except that those pertaining to the stacking gel are omitted. Additional buffers are described by McLellan (1982).

III. Related Methods

A. Determining Protein Molecular Weight

Molecular weight determination is most commonly done by SDS-polyacrylamide gel electrophoresis where proteins are separated primarily by size. For molecular weight determination using nondenaturing gel electrophoresis, the protein sample must be run under a variety of acrylamide concentrations, often ranging from 4% to 12%. The accumulated information from these conditions serves to reduce the effect due to protein charge.

A simplified description of the data treatment follows (for a more detailed version, see Hedrick and Smith, 1968). Protein mobilities are calculated as the R_f value (distance of protein migration divided by distance of migration of the dye front). A semilogarithmic plot of the R_f relative to the acrylamide concentration (Fig. 6.7) should provide a line with a slope characteristic for a protein of a specific molecular weight. Proteins of known molecular weight should be electrophoresed under the same conditions, and the slopes generated from these experiments define a linear relationship with the molecular weight (Fig. 6.8). The molecular weight of the unknown protein of interest may be extrapolated from the data with the molecular weight standards.

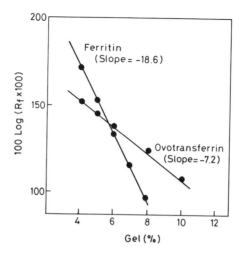

Figure 6.7. Protein relative mobility as a function of acrylamide concentration. The value (100 log(100R$_f$) is plotted versus the gel percentage. Modified from Hedrick and Smith (1968).

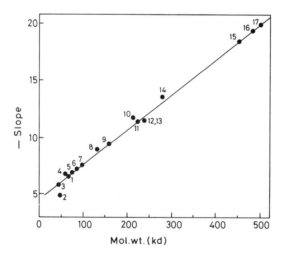

Figure 6.8. Negative slopes from graphs of (100log(100R$_f$)) versus gel percentage plotted as a function of protein molecular weight. Slopes were obtained from graphs such as the one shown in Figure 6.7. The proteins included are 1. pepsin, 2. ovalbumin, 3. α-amylase, 4. albumin, 5. transferrin, 6. ovotransferrin, 7. hexokinase, 8. lactate dehydrogenase, 9. ketose-1-phosphate aldolase, 10. β-amylase, 11. nicotinamide deaminase, 12. α-urease, 13. catalase, 14. xanthine oxidase, 15. apoferritin, 16. urease, and 17. ribulose diphosphate carboxylase. Modified from Hedrick & Smith (1968).

B. Determining Enzyme Activity after Electrophoresis

Enzyme activity may be assayed after polyacrylamide gel electrophoresis either within the gel or following protein elution from the gel. It is not to be assumed that every protein will retain its activity following gel electrophoresis; however, many proteins have been shown to be active after electrophoresis.

Activities have been determined within polyacrylamide gels for a wide variety of enzyme classes. A good reference for experimental details is Gabriel (1971) which describes gel localization of enzymes including methods for dehydrogenases, transferases, hydrolases, lyases, and isomerases.

In order to remove a protein from the gel matrix, it is first necessary to determine the location of the protein in the gel. While staining with Coomassie Blue or silver is common practice for protein visualization, these treatments may denature or modify the protein. An alternative procedure is fluorescence visualization of protein bands as reported by Leibowitz and Wang (1984).

As discussed in Chapter 5 (Section IV.D.), either passive diffusion or electroelution can be used to extract proteins from polyacrylamide gels. For passive diffusion the protein is extracted after cutting the gel in small pieces with a razor and eliminating the gel debris by centrifugation (Bhown and Bennet, 1983). However, the electroelution procedures, such as reverse electroelution (Otto and Snejdarkova, 1981), may provide better yields.

Rothe and Maurer (1986, pp. 55-56) provide a referenced table of over 40 proteins which were separated by polyacrylamide gel electrophoresis and identified by their enzymatic activity. Hubby and Lewontin (1966) describe enzyme assays for use in polyacrylamide gels for esterase, malic dehydrogenase, glucose-6-phosphate dehydrogenase, alkaline phosphatase, α-glycerophosphate dehydrogenase, and leucine aminopeptidase. Similar descriptions for the detection of α-amylase and acetaldehyde oxidase are found in Prakash et al. (1969).

IV. References

Bhown, A.S. and J.C. Bennett. 1983. Meth. Enzymol. 91: 450-455. High-Sensitivity Sequence Analysis of Proteins Recovered from Sodium Dodecyl Sulfate Gels.

Gabriel, O. 1971. Meth. Enzymol. 22: 578-604. Locating Enzymes on Gels.

Hames, B.D. 1981. pp. 23-64 in Gel Electrophoresis of Proteins: A Practical Approach. B.D. Hames and D. Rickwood, eds. 290 pages. IRL Press, London, England.

Hedrick, J.L. and A.J. Smith. 1968. Arch. Biochem. Biophys. 126: 155-164. Size and Charge Isomer Separation and Estimation of Molecular Weights of Proteins by Disc Gel Electrophoresis.

Hubby, J.L. and R.C. Lewontin. 1966. Genetics 54: 577-594. A Molecular Approach to the Study of Genic Heterozygosity in Natural Populations. I. The Number of Alleles at Different Loci in *Drosophila pseudoobscura*.

Leibowitz, M.J. and R.W. Young. 1984. Anal. Biochem. 137: 161-163. Visualization and Elution of Unstained Proteins from Polyacrylamide Gels.

McLellan, T. 1982. Anal. Biochem. 126: 94-99. Electrophoresis Buffers for Polyacrylamide Gels at Various pH.

Otto, M. and M. Snejdarkova. 1981. Anal. Biochem. 111: 111-114. A Simple and Rapid Method for the Quantitative Isolation of Proteins from Polyacrylamide Gels.

Prakash, S., R.C. Lewontin and J.L. Hubby. 1969. Genetics 61: 841-858. A Molecular Approach to the Study of Genic Heterozygosity in Natural Populations. IV. Patterns of Genic Variation in Central, Marginal and Isolated Populations of *Drosophila pseudoobscura*.

Reisfeld, R.A., V.J. Lewis and D.E. Williams. 1962. Nature 195: 281. Disk Electrophoresis of Basic Proteins and Peptides on Polyacrylamide Gels.

Rothe, G.M. and W.D. Maurer. 1986. pp. 37-140 in Gel Electrophoresis of Proteins. M.J. Dunn, ed. 407 pages. IOP Publishing Limited, Bristol, England.

Chapter 7

Isoelectric Focusing and Two-Dimensional Gel Electrophoresis

7

I. Isoelectric Focusing (IEF)
    A. Introduction
    B. Equipment
    C. Preparing Focusing Gel
    D. Sample Preparation and Loading
    E. Running Isoelectric Focusing
    F. Post-Focusing Procedures
    G. Modifications for a Native Isoelectric Focusing Gel
    H. Discussion

II. Two-Dimensional Gel Electrophoresis
    A. Introduction
    B. Equipment
    C. Protocols
    D. Discussion

III. Suppliers

IV. References

I. Isoelectric Focusing (IEF)

A. Introduction

Isoelectric focusing gel electrophoresis is a technique that separates proteins according to their **net charge**. At physiological pH, exposed arginine, histidine, and lysine residues on proteins are generally positively charged, while aspartic acid and glutamic acid residues are generally negatively charged. Thus, at a given pH, a protein's net charge will depend its relative number of positive and negative charges. At a lower pH, the net charge will be more positive, and at a higher pH, the net charge will be more negative.

The pH at which the positive charges on a protein equal the negative charges (in other words, the pH at which the net charge of the protein is zero) defines that protein's **isoelectric point** (pI). During isoelectric focusing, separation is accomplished by placing the protein in a pH gradient generated by an electric field. Under these conditions, the protein migrates until it reaches a position in the pH gradient at which its net charge, or isoelectric point, is zero. Further discussion of protein isoelectric points can be found in Chapter 9, Section I.A.

A powerful addition to isoelectric focusing was first demonstrated by O'Farrell (1975), who placed an isoelectric focusing gel over an SDS-polyacrylamide gel and found that the focused proteins were separated in a second dimension according to their molecular weights. Two-dimensional polyacrylamide gel electrophoresis has opened new possibilities for separating and studying complex protein mixtures (see Section II of this chapter).

For isoelectric focusing, protein bands are found to resolve better when a high voltage gradient is established across the gel. A high voltage gradient can be maintained only if efficient gel cooling is achieved. This requires efficient heat transfer between the gel and the liquid surrounding it. We have chosen to highlight the use of slab gels for isoelectrophoresis rather than the more traditional tube gels to take advantage of the superior heat transfer capabilities of slab gels. In addition, the use of slab gels allows easy comparison of several protein samples.

Since isoelectric focusing is extremely sensitive to charge differences, reproducibility requires that a protein be handled with great care to avoid any modification of the protein's chemical composition or structure during sample preparation. In addition, interactions of proteins with lipids or with other proteins may cause charge modifications which will result in shifted isoelectric mobilities or streaking in the gel. Unless specific protein-protein interactions are being studied or the protein must be maintained in a functional state, it is standard practice to carry out electrophoresis in a denaturing gel system with urea. Further improvements in resolution may be obtained with the use of nonionic detergents.

- Theory: Allen et al. (1984), pp. 63-70.

- Time Required:

    Individual Steps:

Pouring the Gel	90 minutes
Focusing	3 hours
2nd Dimension	
Equilibrating the Gel	30 minutes
Loading the Gel	15 minutes
Electrophoresis	45 minutes
(Not necessary to fix afterwards)	
Fixing the Gel	2 hours - overnight
Coomassie Staining	30 minutes (for major bands)

    Total Time: 7 hours including complete destaining

B. Equipment

- Bio-Rad Mini-Protean II Gel System or other slab mini-gel apparatus
- Power supply (capacity 200 V, 500 mA)
- Hamilton syringe or disposable gel loading tip for micropipette
- Small container for fixing and staining gel

C. Preparing Focusing Gel

- **Reagents**

  1. Acrylamide
  2. Bis-acrylamide
  3. Ampholyte solutions
  4. Urea, sequencing grade
  5. Ammonium persulfate
  6. TEMED
  7. Triton X-100
  8. 2-Mercaptoethanol
  9. Bromophenol blue
  10. Phosphoric acid
  11. Sodium hydroxide (NaOH)
  12. Potassium chloride (KCl)
  13. Trichloroacetic acid (TCA)
  14. Coomassie Blue R-250
  15. Methanol
  16. Acetic Acid

- **Stock Solutions**

  1. Solution A: 30% (w/v) acrylamide, 1% (w/v) bis-acrylamide
     (see Chapter 5)

  2. 20% Triton X-100

  3. 10% Trichloroacetic acid

  4. 1% Trichloroacetic acid

  5. 1% Bromophenol blue

  6. Coomassie Gel Stain (see Chapter 5, Section I.F.)

  7. Coomassie Gel Destain (see Chapter 5, Section I.G.)

• **Ampholyte Selection**

Ampholytes are amphoteric compounds which are provided as a mixture of molecules with closely spaced isoelectric points. They consist of oligo-amino acids and oligo-carboxylic acids in the 600 to 900 dalton molecular weight range (Righetti, 1989).

Giulian et al. (1984) have devised a table of ampholyte blends for use in the preparation of isoelectric focusing gels with different ranges of pH. It is reproduced below:

pH Range	Ampholyte pH Range	% in Final Gel Mixture
pH 3.5-10	pH 3.5-10	2.4%
pH 4-6	pH 3.5-10	0.4%
	pH 4-6	2%
pH 6-9	pH 3.5-10	0.4%
	pH 6-8	1%
	pH 7-9	1%
pH 9-11	pH 3.5-10	0.4%
	pH 9-11	2%

7

- **Pouring the Gel**

  Characteristics: 5% T, 3.3% C (for definition of %T and %C, see Chapter 5, Section I.I.); adapted from Robertson et al. (1987).

  Example of denaturing isoelectric focusing gel preparation:
  (For two 8 cm x 7 cm x 0.75 mm minigels, prepare 12 ml)

  This example is for a gradient from pH 4-6. See Ampholyte Selection section above for recommended ampholyte mix to use for other pH ranges.

  > 5.4 ml H_2O
  > 2.0 ml Solution A
  > 48 µl ampholyte solution, pH 3.5-10
  > 240 µl ampholyte solution, pH 4-6
  > 6. 0g ultrapure urea
  > _____
  > 25 µl 10% ammonium persulfate
  > 20 µl TEMED

  Do not add TEMED until you are at the appropriate point (#4) in the numbered instructions below.

7

1. **Assemble gel plates** according to manufacturer's instructions.

2. **Combine urea, water, Solution A, and ampholyte solution** in a small Erlenmeyer flask. Wear gloves when working with acrylamide, which is a neurotoxin (See comments in Chapter 5).

3. **Mix well** without shaking vigorously. Urea dissolves more rapidly if solution is warmed slightly.

4. **Add ammonium persulfate and TEMED, and mix gently**. Then, **immediately pour acrylamide solution into assembled gel plates**, taking care to pour slowly along a spacer so no bubbles are trapped. Fill gel plates to the rim with acrylamide solution. Polymerization is in progress at this point, so it is important to work rapidly.

5. **Insert comb** so that teeth are entirely surrounded by gel. Be careful not to trap air bubbles in the teeth of the comb.

6. **Allow gels to polymerize** (about 1 hour).

7. After gels have polymerized, **remove comb** carefully.

8. **Attach gels to inner cooling core** and insert into electrophoresis tank. Before loading protein sample, it is useful to add anolyte (10 mM phosphoric acid) to upper buffer chamber to verify that the gel assembly does not leak. To achieve a better seal, petroleum jelly or agarose may be applied to the sealing gasket as well as to the gel plates.

- **Comments**

  1. Remove unpolymerized material at the bottom of the wells or it will polymerize during electrophoresis (Giulian et al., 1984).

  2. Precast gels are commercially available, but only for horizontal slab electrophoresis systems (Pharmacia BioTech, Hoefer).

D. Sample Preparation and Loading (adapted from Robertson et al., 1987)

- **Capacity per Well** (Bio-Rad Mini-Gel System)

Gel Thickness	1 Well	5 Wells	10 Wells	15 Wells
0.5 mm	0.7 ml	45 µl	16 µl	9 µl
0.75 mm	1.0 ml	68 µl	24 µl	14 µl
1.0 mm	1.4 ml	90 µl	32 µl	18 µl
1.5 mm	2.1 ml	135 µl	48 µl	27 µl

- **Denaturing Gel Loading Buffer (2x),** for pH gradient 4-6, 5 ml

2.4 g	Urea (8 M)
20 µl	Ampholyte solution, pH 3.5-10
100 µl	Ampholyte solution, pH 4-6
500 µl	20% Triton X-100 (2%)
50 µl	2-mercaptoethanol (1%)
1.7 ml	Distilled water
200 µl	1% bromophenol blue

Can be stored frozen at -20°C in 0.5 ml aliquots (Pollard, 1984)
Rinse well with 1/4 loading buffer.

- **Steps**

    1. Mix protein sample with an equal volume of 2x Loading Buffer. Before applying sample to gel, spin 5 minutes at 10,000 x g (in Eppendorf centrifuge) to remove aggregated protein which will cause streaking (Sinclair and Rickwood, 1981).

    2. Apply the protein sample into the bottom of the well with a Hamilton syringe or disposable gel loading tip.

- **Comments**

    1. For Coomassie staining, 10 - 30 µg of protein from a crude mixture or 5 - 10 µg of a single protein species per lane are reasonable loading estimates.

    2. If samples do not go into solution readily, it is possible to sonicate them, making sure to keep the temperature below 30°C.

E. Running Isoelectric Focusing

- **Electrophoresis Solutions** (O'Farrell, 1975)

    1. Add catholyte (20 mM sodium hydroxide) to upper buffer chamber.

    2. Add anolyte (10 mM phosphoric acid) to lower buffer chamber.

- **Comments**

    1. Anolyte should be made fresh from a 1 M phosphoric acid stock solution and catholyte should be made fresh from 1 M sodium hydroxide stock solution (Pollard, 1984).

    2. Work at room temperature to prevent precipitation of urea.

- **Focusing Conditions** (from Robertson et al., 1987)

    1. Attach electrodes.

    2. Run for 30 minutes at 150 V (constant voltage).

    3. Then set at 200 V for 2.5 hours (constant voltage). Current will be about 10 mA at the start and will decrease during focusing. Inner chamber will heat to 40 - 50°C during the course of electrophoresis.

F. Post-Focusing Procedures

- **Determining pH Gradient**

   1. Cut a strip of gel into 0.5 cm or 1 cm slices.
   2. Suspend each slice in 1 ml 10 mM KCl for about 30 minutes.
   3. Read pH of KCl solutions.

- **Fixing the Gel** (from Robertson et al., 1987)

   1. Place gels in 10% trichloroacetic acid (TCA) for 10 minutes.
   2. Replace with 1% TCA and soak for at least 2 hours (to remove ampholytes). Soaking overnight is best for reducing Coomassie staining of ampholytes.

- **Staining the Gel**

   1. Stain gel for 10 minutes in Coomassie Gel Stain (see Chapter 5), rocking gently.

   2. Remove Coomassie Gel Stain and replace with Coomassie Gel Destain (see Chapter 5). Replace Destain several times, allowing destaining to continue overnight.

   3. Gel may be dried as described in Chapter 5. A typical gel is presented in Fig. 7.1.

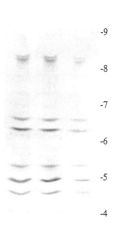

Figure 7.1. Isoelectric focusing gel published by Robertson et al., (1987), kindly provided by H.K. Dannelly.

G. Modifications for a Native Isoelectric Focusing Gel

To run a native isoelectric focusing gel, the following protocol modifications must be made:

- Pouring the Gel (pH 4-6)

  Native Isoelectric Focusing Gel
  5% T, 3.3%C (see Chapter 5, Section I.I for explanation of % T and % C). For two 8 cm x 7 cm x 0.75 mm minigels, need to prepare 12 ml.

  > 9.7 ml H_2O
  > 2 ml Solution A
  > 48 µl ampholyte solution pH 3.5-10
  > 240 µl ampholyte solution pH 4-6
  > _____
  > 50 µl 10% ammonium persulfate
  > 20 µl TEMED

- Sample Preparation and Application

  Native Gel Sample Buffer (2x), 5 ml
  > 3 ml glycerol
  > 200 µl ampholytes (same proportions as for gel)
  > 1.8 ml H_2O

  1. Mix protein sample with an equal volume of 2x Sample Buffer. Spin 5 minutes at 10,000 x g (in Eppendorf centrifuge) before applying sample.

  2. Load sample into well.

- Focusing Conditions (Robertson et al., 1987) -- to be carried out at room temperature.

  1. Attach electrodes.

  2. Set power for 1.5 h at 200 V, then 1.5 h at 400 V (constant voltage).

H. Discussion

- **Troubleshooting** (adapted from Allen et al., 1984)

  1. **Incomplete focusing** is evidenced by fuzzy bands. This may be due to problems in the electrophoresis or to large proteins which have restricted mobility in the gel. If focusing is carried out for too short or too long a period, band resolution is decreased. Increasing the voltage gradient incrementally towards the end of the run sharpens bands. High molecular weight proteins may focus better in more porous agarose gels (see Comments).

  2. **Skewed bands** are usually due to faults in the pH gradient. Verify that the electrodes are clean and make good contact with the gel. Also, be aware of aberrant effects at the edge of the gel.

  3. **Protein band streaking** is a recurrent problem in isoelectric focusing. There are a number of possible causes:

     a. Protein aggregation or precipitation, especially near the protein's isoelectric point, or sample overloading. 8 M urea usually counteracts a tendency to aggregate. Detergents such as Triton X-100 and Nonidet P-40 are useful especially to solve aggregation problems with membrane proteins (see Comments). Be sure to centrifuge samples before loading gel to remove particulates.

     b. Presence of nucleic acids in the sample. Various procedures to remove nucleic acids, including acid extraction, salt precipitation, and nuclease treatment, are cited in Sinclair and Rickwood (1981, p. 194).

     c. Protein modification. When the urea is not ultra-pure, isocyanate impurities may cause protein carbamoylation. Prerunning the gel can remove isocyanate. Other modifications include oxidation of Cys residues or deamination of Asn or Gln when the protein sample is improperly handled and stored.

  4. **Wavy bands** are often due to a high salt content of the samples. Sometimes wavy bands are attributed to impurities in the ampholyte or electrolyte solutions, or to dirty electrodes.

5. **Uneven pH gradient** may result from electrode contact not being parallel to the gel, impurities within the gel, or ampholyte concentrations which are too low. If the pH gradient in the alkaline portion of the gel is lost, this is likely due to cathodic drift. Supplement with pH 9-11 ampholytes or run a nonequilibrium pH gradient electrophoresis gel (NEPHGE, see Comments).

6. **High background staining** is probably due to ampholytes remaining in the gel after fixation. Increase the time of fixing with 1% TCA.

7. **Missing or faint bands** are likely to be low molecular weight proteins (<10 kd) or proteins which have not been denatured during fixation. Increase TCA concentration or fix with glutaraldehyde (see Chapter 5).

8. **Overlapping spots** may occur in complex protein mixtures. Changing the pH range of the isoelectric focusing gel may solve this problem. Further steps of protein purification or immunoprecipitation are recommended to remove an overlapping spot.

7

- **Comments**

    1. To establish the position of the dye front after isoelectric focusing, mark the dye front with a fine gauge wire (0.1 mm diameter). The wire may be dipped in India ink for increased visibility.

    2. Different brands of ampholytes have slightly different properties, and gels run with ampholytes from different sources will have differing patterns of protein separation. For the best reproducibility, do not change ampholyte brands.

    3. In general, shallower pH gradients will lead to better resolution of protein bands. However, shallower pH gradients require a longer focusing time. A good compromise is a pH range of 2 (der Lan and Chrambach, 1981).

    4. Because of the limitations of power supplies and gel cooling, the maximum recommended length for gels is 8 - 10 cm. It is better to run several gels with pH gradients than a single long gel (Allen et al., 1984).

    5. Cathodic drift becomes a problem when gels are subjected to electrophoresis for very long periods (>3000 volt-hours) due to lability of the ampholytes. Cathodic drift results in the partial collapse of the pH gradient, particularly at pH values above 8. To overcome this problem, O'Farrell et al. (1977) developed a technique in which isoelectric focusing gels were run for shorter periods (1600 volt-hours). This method permits focusing of proteins at higher pH ranges and is called nonequilibrium pH gradient electrophoresis (NEPHGE). However, it is not possible to determine a protein's isoelectric point using NEPHGE. See O'Farrell et al. (1977) or Phillips (1988) for instructions.

    6. Urea solubilizes proteins and eliminates protein-protein and protein-lipid interactions. This allows separation to occur more rapidly and improves resolution. Urea also reduces (but does not eliminate) cathodic drift. It should be noted that the isoelectric point of a denatured protein may differ from that of the native protein.

7. If a protein solution contains SDS, add urea to the sample to a final concentration of 8 M. At high concentrations of urea, SDS interacts minimally with proteins.

8. 2% Nonidet P-40 or 2% Triton X-100 (O'Farrell et al., 1977; Giulian et al., 1984) are often added to denaturing solutions to help keep proteins (especially membrane proteins) soluble. Some authors recommend zwitterionic detergents such as CHAPS and Zwittergent 3-14.

9. Ultra-thin gels (50 - 500 μm thick) provide several advantages over standard slab gel isoelectric focusing. The thinness of the gel allows increased speed and resolution because of higher field strengths and improved cooling. Protocols may be found in Allen et al. (1984), Giulian et al. (1984), and Radola (1983).

10. Ready-made Immobiline isoelectric focusing gels (Pharmacia-LKB), in which the ampholytes are covalently bonded to the acrylamide, are available for high-resolution separation in the 4-7 pH range.

11. High molecular weight proteins (> 750 kd) often have aberrant mobility in polyacrylamide isoelectric focusing gels. Agarose or agarose-acrylamide gels have provided satisfactory alternatives. For protocol descriptions, see Allen et al. (1984) and Pino and Hart (1984).

12. Coomassie staining may be improved if the TCA fixation step is followed by a 10 - 30 minute rinse of the gel in a solution of 0.25% SDS in ethanol:acetic acid:water (33:10:57). SDS binds ampholytes, leading to improved removal of the ampholytes from the gel (Giulian et al., 1984).

II. Two-dimensional Gel Electrophoresis

A. Introduction

The second dimension of two-dimensional gel electrophoresis simply consists of SDS-polyacrylamide gel electrophoresis as described in Chapter 5. A strip of gel or a tube gel from isoelectric focusing (the first dimension) is fitted over an SDS-polyacrylamide gel and the proteins are separated according to molecular weight by electrophoresis. Pre-equilibration of the isoelectric focusing gel in SDS is necessary prior to running the second dimension. Although no special equipment is needed, a number of two-dimensional gel electrophoresis devices are commercially available, included a 2-D tube gel accessory for the Bio-Rad Mini-Protean II Gel Electrophoresis cell.

B. Equipment

- Bio-Rad Mini-Protean II Gel Electrophoresis Cell

- Power supply (capacity 200 V, 500 mA)

- Container for staining and destaining the gel

C. Protocols

- Reagents

  Glycerol
  2-Mercaptoethanol
  Sodium dodecyl sulfate (SDS)
  Tris
  For gel electrophoresis reagents, consult Chapter 5, Section I.C.

- Working Solutions

  1. 20% SDS
  2. 1 M Tris-HCl (pH 6.8)

- Steps

    1. If the second dimension gel is to be run immediately following isoelectric focusing, pour a 1.0 or 1.5 mm SDS-polyacrylamide gel at the same time as pouring the isoelectric focusing gel. Instead of placing a comb into the stacking gel, leave about 0.5 cm above the stacking gel and overlay very carefully with water. The isoelectric focusing gel may be stored frozen and run in the second dimension later (see below).

    2. Run isoelectric focusing gel as described in the first part of this chapter.

    3. After focusing, cut a strip of gel 0.5 cm wide and place in a small container or a boat made of parafilm. Hint: If possible, use a 5-well comb to give very wide lanes, then cut a 0.5 cm strip down the center of the desired lane.

    4. Equilibration

        Equilibration Buffer, 100ml
        5 ml 2-Mercaptoethanol   (5%)
        6.25 ml 1 M Tris-HCl (pH 6.8)   (62.5 mM)
        11.5 ml 20% SDS   (2.3%)
        10 ml Glycerol   (10%)
        H_2O to 100 ml

        a. Add equilibration buffer to IEF gel.
        b. Incubate for 15 - 30 minutes.
        c. Gel can be immediately loaded onto an SDS-polyacrylamide gel or frozen and stored at -80°C.

5. Electrophoresis

   a. Remove water overlay from stacking gel by aspiration.
   b. Overlay the stacking gel with electrode buffer (see Chapter 5 for preparation details).
   c. Wearing gloves and being careful not to distort the gel, place the isoelectric focusing gel strip on the stacking gel. It is a good idea to lower the gel strip gently onto the stacking gel from one side with a spatula or syringe to keep from trapping air bubbles. When loading gel, note which side is acidic and which is basic from isoelectric focusing.
   d. Proceed with electrophoresis and staining as described in Chapter 5 (Sections I. F., G., and H.). An example of a two-dimensional gel is presented in Fig. 7.2.

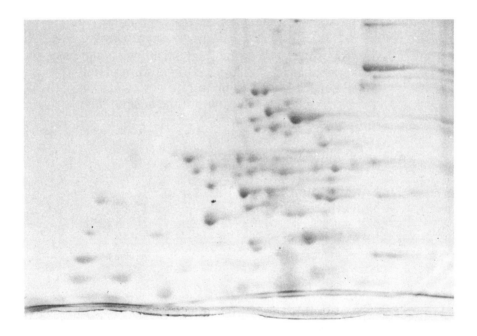

Figure 7.2. Two-dimensional gel of a crude extract of *Escherichia coli* stained with Coomassie Blue. Separation in the horizontal dimension was achieved by isoelectric focusing in the pH range 4 - 7 in the presence of 8 M urea, followed by separation in the vertical dimension by SDS-polyacrylamide gel electrophoresis. From Robertson et al. (1987), kindly provided by H.K. Dannelly.

D. Discussion

- A stacking gel sometimes causes band elongation in the direction of isoelectric focusing due to lateral band spreading (Strahler et al., 1989). If this is a problem, the stacking gel may be omitted.

- Streaking in second dimension (SDS-PAGE) may be due to poor equilibration of the first dimension gel or insufficient buffering capacity of the Tris in the second dimension. 2-Mercaptoethanol may cause streaks with silver staining (see Chapter 5).

- To run molecular weight markers in the second dimension, embed the markers in a 1% agarose plug with 0.02% bromophenol blue. The plug should be poured in a tube of the right diameter, and then the solidified gel is cut into 0.5 - 1.0 cm sections which can be stored frozen. 5 - 10 µg of molecular weight proteins per slice are adequate. The agarose plug is inserted into one end of the SDS-polyacrylamide gel prior to electrophoresis.

- Ampholytes run as small proteins in the gel, are acid precipitable and will stain. Elution of ampholytes during the fixation step following focusing is essential for removal of background staining; however, extended incubation at this step may cause protein elution from the gel. Protein elution during fixation may make reproducibility difficult to achieve.

- Alternative two-dimensional gel systems are used for separation of special protein classes. Sinclair and Rickwood (1981) discuss systems for separation of ribosomal proteins, histones, and other nuclear proteins (pp. 209-217).

- Isoelectric focusing may be used on a preparative scale for protein isolation. Since the same principles are employed, similar problems are encountered as for analytical isoelectric focusing. In general, three kinds of matrices have been employed: polyacrylamide gel, agarose, and granulated gels. For laboratory-scale separations, one gram of material is considered the upper limit for separation, and this amount may be separated in under an hour with certain systems. For a good description of the method, see Radola, 1984. For protocols, consult Radola (1984), Allen et al. (1984), or der Lan and Chrambach (1981).

III. Suppliers

Ampholyte solutions:
- BioRad: BioLytes
- Pharmacia: Ampholines, Immobilines, Pharmalytes
- Serva: Servalyts

Chemicals for gel electrophoresis: Aldrich, BDH, BioRad, Calbiochem, Fluka, Kodak, Merck, Pharmacia, Sigma, Whatman

Electrophoresis Cell:
- BioRad Mini-Protean II Electrophoresis Cell
- Hoefer Mighty Small II System
- Pharmacia Midget System

- Commercial apparatus for ultra-thin and preparative isoelectric focusing are available from BioRad, Hoefer, and Pharmacia.

Hamilton Syringes: Hamilton

Power Supply:
- BioRad Model 250/2.5 Power Supply
- Hoefer PS500XT or PS500X Power Supply
- Pharmacia EPS 500/400 or GPS 200/400 Power Supply

IV. References

Allen, R.C., C.A. Saravis and H.R. Maurer. 1984. pp. 63-147,
Isoelectric Focusing and pp. 148-180, Multiparameter Techniques.
In Gel Electrophoresis and Isoelectric Focusing of Proteins:
Selected Techniques. 255 pages. Walter de Gruyter, Berlin.

Giulian, G.G., R.L. Moss and M. Greaser. 1984. Anal. Biochem. 142:
421-436. Analytical Isoelectric Focusing Using a High-Voltage
Vertical Slab Polyacrylamide Gel System.

der Lan, B. and A. Chrambach. 1981. pp. 157-188 in Gel
Electrophoresis of Proteins: A Practical Approach. Hames, B.D.
and D. Rickwood, eds. 290 pages. IRL Press, Oxford and
Washington, D.C.

O'Farrell, P.H. 1975. J. Biol. Chem. 250: 4007-4021. High
Resolution Two-Dimensional Electrophoresis of Proteins.

O'Farrell, P.Z., H.M. Goodman, and P.H. O'Farrell. 1977. Cell 12:
1133-1142. High Resolution Two-Dimensional Electrophoresis of
Basic as Well as Acidic Proteins.

Phillips, T.A. 1988. DNA and Protein Engineering Techniques 1: 5-9.
Two-Dimensional Polyacrylamide Gel Electrophoresis of Proteins.
Alan R. Liss, New York.

Pino, R.M. and T.K. Hart. 1984. Anal. Biochem. 139: 77-81.
Isoelectric Focusing in Polyacrylamide-Agarose.

Pollard, J.W. 1984. pp. 81-96. Two-Dimensional Polyacrylamide
Gel Electrophoresis of Proteins. pp. 81-96 in Methods in
Molecular Biology, Vol. 1, Proteins, J.M. Walker, ed. 365 pages.
Humana Press, Clifton, New Jersey.

Radola, B.J. 1983. pp. 101-118. Ultra-Thin-Layer Isoelectric
Focusing. In Electrophoretic Techniques, C.F. Simpson and M.
Whittaker, eds. 280 pages. Academic Press, London.

Radola, B.J. 1984. Meth. Enzymol. 104: 256-274. High-Resolution
Preparative Isoelectric Focusing.

Righetti, P.G. 1989. pp. 23-63 in Protein Structure: A Practical
Approach. T.E. Creighton, ed. 355 pages. IRL Press, Oxford.

Robertson, E.F., H.K. Dannelly, P.J. Malloy and H.C. Reeves. 1987.
Anal Biochem. 167: 290-294. Rapid Isoelectric Focusing in a
Vertical Polyacrylamide Minigel System.

Sinclair, J. and D. Rickwood. 1981. Two-Dimensional Gel
Electrophoresis. pp. 189-218 in Gel Electrophoresis of Proteins: A
Practical Approach, B.D. Hames and D. Rickwood, eds. 290
pages. IRL Press, Oxford and Washington, D.C.

Strahler, J.R., R. Kuick and S.M. Hanash. pp. 65-92 in Protein
Structure: A Practical Approach. T.E. Creighton, ed. 355 pages.
IRL Press, Oxford.

Chapter 8

Immunoblotting

I. Introduction

II. Performing an Immunoblot
    A. Equipment
    B. Reagents
    C. Protocols
    D. Protocol Modifications for Other Detection Methods
    E. Staining for Total Protein
    F. Erasing Immunoblots

III. Discussion
    A. Membrane Storage
    B. Transfer Anomalies
    C. References for Other Uses of Immunoblots

IV. Suppliers

V. References

8

I. Introduction

This chapter describes immunochemical techniques, termed immunoblotting or Western blotting, which are used to detect a protein immobilized on a matrix (Towbin et al., 1979). Before employing this procedure, it is necessary to have a monoclonal or polyclonal antibody capable of recognizing the protein of interest and a solution containing the protein, either a crude extract or a more purified preparation. Immunoblotting is an extremely powerful technique for identifying a single protein (or epitope) in a complex mixture following separation based on its molecular weight (SDS-PAGE, Chapter 5), size and charge (nondenaturing gel electrophoresis, Chapter 6) or isoelectric point (isoelectric focusing, Chapter 7). In addition, immunoblotting combined with immunoprecipitation permits the quantitative analysis of minor antigens. A number of other applications for immunoblots are listed below. The immobilization of proteins on a membrane matrix is preferred to working directly with a polyacrylamide gel because the proteins are more accessible, membranes are easier to handle than gels, smaller amounts of reagents are needed, and processing times are shorter (Gershoni and Palade, 1982).

Limits of Detection: 10 pg (picogram = 10^{-12} g) with horseradish peroxidase or alkaline phosphatase labeling or 1 pg with immunogold or ^{125}I labeling.

Theory: Gershoni and Palade, 1983

Description of Immunoblotting

Immunoblotting can be divided into two steps: transfer of the protein from the gel to the matrix, and decoration of the epitope with the specific antibody.

Protein transfer is most commonly accomplished by electrophoresis. The two common electrophoretic methods are:

1. Semi-dry blotting, in which the gel and immobilizing matrix are sandwiched between buffer-wetted filter papers through which a current is applied for 10 - 30 minutes.

2. Wet (tank) blotting, in which the gel-matrix sandwich is submerged in transfer buffer for electrotransfer, which may take as little as 45 minutes or may be allowed to continue overnight.

We present only wet blotting here, since it permits greater flexibility without being significantly more expensive in time or materials. The Bio-Rad Mini Trans-Blot transfer cell is used for describing the electrotransfer, but the conditions should be easily adaptable to other electroblotting devices.

Following transfer, detection of the epitope proceeds in two or three steps. First, the nitrocellulose membrane is incubated with the primary antibody for several hours or overnight. Next, a second antibody (Protein A may also be used), which recognizes an epitope on the first antibody is incubated with the nitrocellulose. Typically, the second antibody (usually goat antibodies raised against rabbit immunoglobulins for rabbit-generated first antibodies) is purchased already conjugated to a labeling agent such as the enzyme horseradish peroxidase. This marker is then visualized by a colorimetric reaction catalyzed by the enzyme which yields a colored product that remains fixed to the nitrocellulose membrane. Other detection systems including alkaline phosphatase or immunogold conjugates and ^{125}I labels are also described below.

Blotting Membranes

Two kinds of membranes are most commonly used for transferring proteins from gels: nitrocellulose and nylon. For most applications, nitrocellulose appears to be the membrane of choice and this chapter will describe this support only for protein immobilization. Nitrocellulose is preferred because it is relatively inexpensive and blocking nitrocellulose from nonspecific antibody binding is fast and simple.

Nylon may be more useful if **1.** a higher protein binding capacity is required (480 $\mu g/cm^2$ vs. 80 $\mu g/cm^2$ for nitrocellulose; Gershoni and Palade, 1982); **2.** the protein to be studied binds weakly to nitrocellulose (especially high molecular weight or acidic proteins); or **3.** greater resistance to mechanical stress is desired. The use of nylon has been limited because it is more expensive than nitrocellulose, blocking is cumbersome, and staining for total protein with anionic dyes is not possible.

Other Applications Involving Immobilized Proteins (see Discussion)

1. Epitope mapping
2. Structural domain analysis
3. Dot blotting
4. Renaturing proteins for functional assay
5. Ligand binding
6. Improved autoradiography
7. Antibody purification
8. Cutting out protein bands for antibody production
9. Protein identification: amino acid analysis and protein sequencing

8

II. Performing an Immunoblot

A. Equipment

- Electroblotting Apparatus
  We describe the Bio-Rad Mini Trans-Blot Electrophoretic Transfer Cell. Other commercially available or homemade apparatus can also be used.

- Power supply with capacity of 200 V, 0.6 A

- Whatman 3MM paper

- Nitrocellulose paper, 0.2 or 0.45 µm pore size
  Nitrocellulose should be stored in a cool dark place and gloves should be worn when handling the membrane to prevent protein transfer.

- Heat-sealable plastic bags and sealer (e.g. Seal-A-Meal or Kapak)

- Small plastic or glass container for gel incubations (15 x 10 x 2 cm)

- Rocker or rotary shaker

- Shallow tray used in preparing gel for transfer (30 x 15 x 3 cm)

- Magnetic stir plate

8

B. Reagents

- **Transfer Buffer:**
  1. Tris Base
  2. Glycine

- **Tris Buffered Saline (TBS):**
  1. Tris-HCl, 2 M solution at pH 7.5, for instructions see
     Chapter 5 section I.C.
  2. Sodium chloride (NaCl), 4 M solution

- **Blocking Solution**: Bovine serum albumin (BSA)

- **Horseradish Peroxidase Developing Reagent:**
  1. Chloronaphthol
  2. Methanol
  3. Hydrogen peroxide (H_2O_2)

- **Alkaline Phosphatase Developing Reagent:**
  1. $MgCl_2$
  2. 5-Bromo-4-chloro-3-indolyl phosphate (BCIP)
  3. Dimethylformamide (DMF)
  4. Veronal acetate buffer
  5. *p*-Nitro blue tetrazolium chloride (NBT)

- **Immunogold Incubation Buffer**: Tween 20

- **Amido Black Total Protein Stain:**
  1. Amido Black 10B (Naphthol Blue Black, Buffalo Black)
  2. Isopropanol
  3. Acetic acid

- **India Ink Total Protein Stain:**
  1. Tween 20
  2. India Ink

- **Immunoblot Erasing Buffer:**
  1. Powdered nonfat milk
  2. Sodium dodecyl sulfate (SDS)
  3. Tris-HCl
  4. 2-Mercaptoethanol

C. Protocols

• **Transfer**

Time required: minimum of 90 minutes, or overnight

This protocol begins once a polyacrylamide gel (SDS, native or IEF) has already been run to separate the protein of interest from other proteins in the sample. It is recommended to run two such gels, staining one directly and using the other for immunoblotting. Alternatively, it is possible to stain the nitrocellulose membrane for total protein following electroblotting (see Section E).

Transfer Buffer, 1 liter
    1.93 g Tris Base 15.6 mM
    9 g Glycine 120 mM
Add distilled water to 1 liter, pH should be between 8.1 and 8.4
    without adjustment.
Can also be made up as a 20x stock solution.

Preparations:
    - Make up 1 liter transfer buffer and chill to 4°C.
    - Put distilled water in cooling unit of Trans-Blot Cell and freeze
      overnight.

1. **Prepare transfer cell and nitrocellulose**:
    a. Soak nitrocellulose sheet (6 cm x 8 cm) in transfer buffer (30 -
       40 ml) for 15 - 20 minutes.
    b. Rinse buffer chamber with distilled water.
    c. Insert Trans-Blot electrode insert and small stir bar into buffer
       chamber, and add transfer buffer until half full (about 400 ml).
    d. Insert frozen cooling unit.

2. **Arrange polyacrylamide gel-membrane sandwich** (Fig. 8.1 and 8.2):

   a. In a shallow tray, open the transfer cassette and place a wetted sheet of Whatman 3MM paper (8 x 10 cm) on a well-soaked fiber pad on the gray panel of the transfer cassette. Transfer buffer should be used for soaking.

   b. Carefully place the gel on the wet filter paper on gray side of assembly and arrange well ears and gel so that all air bubbles are removed. Use transfer buffer for lubrication and be sure to wear gloves (Fig. 8.3).

   c. After wetting gel, carefully lay a wetted sheet of nitrocellulose on top, beginning from one side so air bubbles are removed to the edge of the gel (Fig. 8.4). Be sure to wear gloves or use tweezers when handling membranes.

   d. Place a wetted sheet of 3MM paper over the nitrocellulose and roll a small test tube or pipette over the sandwich like a rolling pin to remove any air bubbles (Fig. 8.5).

   e. Cover with the second well-soaked fiber pad, close the transfer cassette and slide it into the electrode insert in the buffer tank, keeping the gray panel of the cassette on the same side as the gray panel of the electrode assembly (Fig. 8.6).

   f. Fill the buffer tank with transfer buffer.

3. **Electrotransfer**:

   a. Place entire Trans-Blot apparatus on a magnetic stir plate and begin stirring.

   b. Attach the electrodes.

   c. Set the power supply to 100 V (constant voltage) and transfer for 1 h. For chilled transfer buffer, initial current will be 0.2 A and final current will be 0.4 A.

- Overnight transfers should be run at 30 V (Bio-Rad Trans-Blot Instruction Manual).

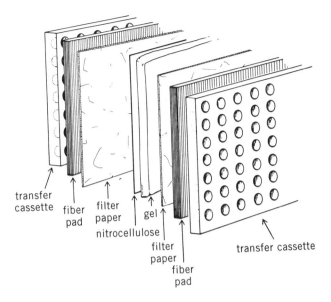

Figure 8.1. Sandwich involving polyacrylamide gel and nitrocellulose membrane for electroblotting.

Figure 8.2. Side view of sandwich in Fig. 8.1.

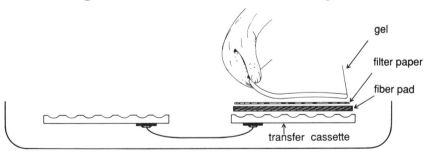

Figure 8.3. Placing the gel on filter paper.

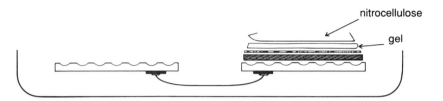

Figure 8.4. Placing nitrocellulose on gel.

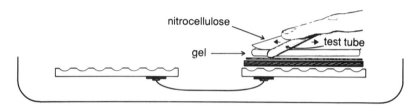

Figure 8.5. Removing air bubbles with a test tube.

Figure 8.6. Inserting transfer cassette into electrode assembly.

- **Comments on Transfer**

    1. Be sure to check the current at the start of transfer. An unusually high current reading is most likely due to improper preparation of the transfer buffer, and will create problems due to excess heat generation during the transfer period.

    2. Lower buffer ionic strength allows higher voltage without high current and heat generation (Bers and Garfin, 1985), but during transfer electrolytes elute from the gel, increasing buffer conductivity and decreasing the resistance (Gershoni and Palade, 1983).

    3. Constant current transfer can be accomplished at 200 mA for two hours (Gershoni and Palade, 1982).

    4. It may be helpful to incubate the gel in transfer buffer for 15 - 30 minutes prior to transfer if distortion of bands due to gel swelling during transfer is a problem or if electrolytes eluting from the gel cause an excessive increase in temperature during the transfer (Gershoni and Palade, 1983).

    5. Using prestained protein molecular weight markers and staining the gel after electroblotting allows visual evaluation of the completeness of transfer.

    6. For large proteins, a lower acrylamide concentration may result in better transfer (Peluso and Rosenberg, 1987). Another possibility is a two-step elution procedure which allows transfer of low and high molecular weight proteins (Otter et al., 1987).

    7. For a guide to transfer conditions for different membranes or transfer buffers, see Bio-Rad Trans-Blot Instruction Manual.

8. The most common problem encountered with electrophoretic transfer of proteins to a membrane is poor protein transfer. This is easily monitored by staining the gel or membrane for proteins after electrotransfer. Four strategies can be tried if the protein of interest binds poorly to nitrocellulose:

   a. Add methanol to 20% in the transfer buffer. Methanol:
     - Decreases efficiency of protein elution from the gel but improves absorption to nitrocellulose (Gershoni and Palade, 1982).
     - Lengthens elution time for large proteins (Gershoni and Palade, 1983) and, due to removal of SDS, can reduce transferability of proteins with a high isoelectric point (Peluso and Rosenberg, 1987 and Bers and Garfin, 1985).
     - Prevents gel swelling (Bers and Garfin, 1985).

   b. Add SDS to 0.1% in the transfer buffer. Low SDS concentrations (around 0.1%) in the transfer buffer can improve transfer efficiency (Towbin and Gordon, 1984), but may reduce the membrane's ability to retain some proteins (Peluso and Rosenberg, 1987).

   c. Use lower pore size nitrocellulose (0.2 μm).

   d. Immerse the filter in glutaraldehyde immediately after transfer. Glutaraldehyde cross-linking of proteins improves the stability of small acidic proteins for immunoblotting which may otherwise diffuse out from the membrane. Use 0.2% glutaraldehyde in TBS for 45 minutes (Van Eldik and Wolchok, 1984).

• **Immunodetection**

Once the proteins have been transferred from the polyacrylamide gel to the nitrocellulose membrane, detection of specific proteins proceeds by the use of antibodies. Prior to the addition of antibodies, the membrane is coated with a blocking agent, typically a 3% solution of bovine serum albumin (BSA) in Tris-buffered saline (TBS). Blocking the membrane is important so that antibodies do not bind nonspecifically to the membrane. The first antibody (also called the primary antibody) recognizes the protein of interest while the second antibody recognizes the F_c portion of the first antibody. The second antibody is coupled to an enzyme or another detectable reagent which produces a colored product. The protocol below describes the use of a horseradish peroxidase-conjugated second antibody.   Subsequently, modifications of the protocol are provided which describe the use of alkaline phosphatase-, gold-, or ^{125}I-conjugated second antibodies. In addition, immunodetection kits are commercially available (Bio-Rad, Pierce).

Sensitivity:
> **Horseradish peroxidase**: 10 - 20 pg (Bers and Garfin, 1985)
> **Alkaline phosphatase**: 10 - 50 pg (Bio-Rad Bulletin 1310, 1987)
> **Immunogold**: 1 - 25 pg (Bio-Rad Bulletin 1310, 1987)
> [125]**I**: 50 - 100 pg, 1 pg with high specific activity [125]I (Bers and
>   Garfin, 1985)

Time required: 4 - 5 hours, or incubations may continue overnight.

Solutions to prepare:
> - TBS (Tris-buffered saline)
> - 3% BSA (Bovine serum albumin) in TBS or 1% nonfat dry milk
>   in TBS
> - 0.5% BSA/TBS

TBS (T̲ris-B̲uffered S̲aline), 1 liter

> 5 ml 2 M Tris-HCl (pH 7.5)10 mM
> 37.5 ml 4 M NaCl 150 mM
> 957.5 ml distilled water
Can also be made as a 10x solution.

All steps can be carried out at room temperature or at 4°C.

1. **Block membrane:**
   a. Disconnect transfer apparatus, remove transfer cassette, and peel 3MM paper from nitrocellulose.
   b. Using forceps or wearing gloves, remove nitrocellulose membrane from transfer apparatus to a small container or to a heat-sealable plastic bag.
   c. Add at least 8 ml 3% BSA/TBS (enough to cover the membrane).
   d. Rock the filter gently for 30 minutes to 1 hour, making sure that the entire filter is in contact with the BSA solution. The blocked membrane may be stored in the BSA/TBS solution with 1 mM NaN$_3$ overnight or for as long as several days depending on the stability of the protein.

2. **Wash membrane:** Pour off BSA solution and rinse briefly with TBS three times.

3. **First antibody wash:**
   a. Pour off TBS, add first antibody at appropriate dilution in 8ml 0.5% BSA/TBS (see comment 2.d. below for dilutions).
   To conserve volume, place the membrane in a heat-sealable plastic bag, add first antibody solution and seal. Using such a bag, it is possible to use a 4 ml of solution for a 6 cm x 8 cm membrane. Be careful to remove air bubbles as completely as possible. Air bubbles prevent antibody-epitope contact and can result in bands that remain undetected.
   b. Rock gently for at least 1 hour. Overnight incubations are possible and may increase detection sensitivity.

4. **Wash membrane:**
   a. Pour off first antibody solution from membrane.
   b. Wash twice for 10 minutes with TBS.

5. **Second antibody wash:**
   a. Pour off TBS.
   b. Add second antibody at appropriate dilution in 8 ml 0.5% BSA/TBS.
   c. Rock the membrane gently for at least an hour. An overnight incubation is acceptable.

6. **Wash membrane:**
   a. Pour off second antibody solution from membrane.
   b. Rinse for 30 minutes with TBS, with 3 changes. At this time, if you are incubating the membrane in a heat-sealable plastic bag, it is advisable to move the membrane to a glass or plastic container.

7. **Develop nitrocellulose** (horseradish peroxidase):
   a. Prepare developing reagent:
      1 ml chloronaphthol solution (30 mg/ml in methanol)
      Add 10 ml methanol
      Add room temperature TBS to 50 ml
      Add 30 µl 30% hydrogen peroxide (H_2O_2)
   b. Pour off TBS from membrane, add developing reagent.
   c. Rock nitrocellulose gently, monitoring development.
   d. If a heavy precipitate forms, replace developing reagent with a freshly made solution.
   e. Development should be complete in 5 to 30 minutes.
   f. Stop development by washing membrane with distilled water for 30 minutes with 3 changes.
   g. Dry nitrocellulose with absorbent paper and store the membrane protected from light and the atmosphere. A plastic envelope in a notebook is usually suitable.
   h. Photograph within a week, since the signal may fade with time and the nitrocellulose may begin to turn yellow. A representative pattern is presented in Fig. 8.7.

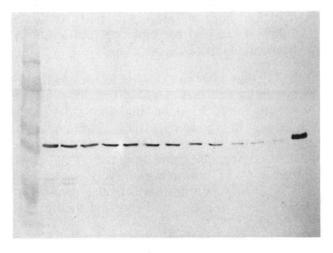

Figure 8.7. Immunoblot detecting yeast tubulin with a tubulin-specific first antibody and subsequent staining with alkaline phosphatase. Pairs of samples were run in a 2-fold dilution series from 20 µg to 0.62 µg. The far left lane contains molecular weight markers and the far right lane contains a brain tubulin marker.

- **Comments on Immunodetection**

1. Membrane Blocking and Washing

a. Nitrocellulose may be baked for 1 hour at 80°C in a vacuum oven following transfer to help stabilize protein binding (Hsu, 1984).

b. In step 1, blocking can also be accomplished with 3% gelatin in TBS. Blocking of nylon membranes should be done with 10% BSA at 45 - 50°C for at least 12 hours.

c. After step 2, the filter may be dried and stored. Rewet filter in 0.5% BSA/TBS before continuing with immunodetection protocol.

d. Bovine serum albumin (BSA) is most commonly used to reduce background in the blocking and washing steps. BSA is often contaminated with immunoglobulin G, which makes it unsatisfactory for use with protein A (Bers and Garfin, 1985). Use Tween-20 or gelatin instead.

e. Other blocking agents include:
  - BLOTTO: 5% (w/v) nonfat dry milk, 0.01% Antifoam A, Sigma (Johnson et al., 1984).
  - 3% gelatin for blocking, 1% for antibody incubations (Bio-Rad Immun-Blot (GAR-HRP) Assay Kit Bulletin).
  - 0.05% Tween-20 for blocking and antibody incubations (Bers and Garfin, 1985).

f. BLOTTO may give inconsistent results with different batches of dry milk, and it is possible that blocking with milk will give a higher background than with BSA (Kaufmann et al., 1987). In addition, these products may contain immunoglobulin G which will become a problem if protein A is used for detection, and they may react with lectin probes such as concanavalin A (Bers and Garfin, 1985). BLOTTO may be omitted from the washing steps if it has already been used for blocking the membrane (Peluso and Rosenberg, 1987).

g. Congealing of gelatin at low temperatures may be a problem (Bers and Garfin, 1985).

2. Use of Antibodies

a. Repeated freezing and thawing of antibodies can lead to aggregation and loss of activity. For routine use, prepare convenient aliquots of the first and second antibodies (steps 3 and 5) and store them in the freezer. Antibodies can be diluted to a convenient concentration in 0.5% BSA/TBS and stored at -20°C (Burnette, 1981).

b. Color development is strongest at high first and second antibody concentrations. However, if antibody concentrations are too high, nonspecific background bands will appear unless the enzyme reaction is stopped promptly. See Bers and Garfin (1985) for additional troubleshooting advice.

c. Antibody dilution guidelines:
   - First antibody: extremely variable, from 1:10 to 1:100,000.
   - Second antibody: commonly 1:500 to 1:4000.

3. General Immunodetection Comments

a. Protein transfer is not the same on both sides of the nitrocellulose membrane (as becomes evident if using pre-stained molecular weight markers). Be sure to look at the signal on both sides of the immunoblot during development.

b. SDS-PAGE may cause a loss of antigenicity, especially for monoclonal antibody detection (Burnette, 1981).

c. Possible reasons for poor color development:
   - First or second antibody is inactive or non-saturating due to improper storage or excess dilution. A control for first antibody activity may involve spotting the antigen directly on a piece of nitrocellulose and performing the immunoassay. Similarly, the second antibody can be tested for binding to a different first antibody produced in the same species.
   - Insufficient antigen on the nitrocellulose membrane. Stain the nitrocellulose for total protein or include a known amount of control antigen on the blot.

d. Possible reasons for high background:
- Insufficient washing between antibody incubations.
- Insufficient blocking.
- Contaminated fiber pads or transfer buffer.
- Antibody concentrations too high. Include 0.05% Tween 20 in antibody buffers.

Additional suggestions for solving problems due to diffuse or specific background bands may be found in Harlow and Lane (1988, p. 510).

4. Horseradish Peroxidase

a. Horseradish peroxidase-coupled second antibody yields a blue-purple color when reacted with chloronaphthol as described above. Alternative development substrates include aminoethylcarbazole (yielding a red color) and diaminobenzidine (yielding a brown color) which may provide greater sensitivity (see Harlow and Lane, 1988).

b. After transferring the second antibody aliquot to the BSA/TBS solution in step 5, keep the tube containing a trace of the second antibody for testing in step 7. Prepare the developing reagent and test the reagent by adding 50 - 100 μl to the used second antibody tube. If the solution does not start to turn blue within 2 minutes, there is a problem with the developing reagent.

c. Chloronaphthol stock solution (30 mg/ml in methanol) can be made and stored at -20°C for at least a year.

d. Although the horseradish peroxidase detection system is economical and rapid to perform, some possible disadvantages include:
- Fading with exposure to light
- Non-specific reaction by endogenous peroxidase enzymes on immunoblot

e. Possible reasons for poor color development:
- Developing reagent is inactive, especially if H_2O_2 is inactive or chloronaphthol has precipitated out of solution. Prepare new developing reagent, using fresh enzyme immunoassay (EIA) grade chemicals.
- Horseradish peroxidase may be inactivated by azide or impure methanol.

D. Protocol Modifications for Other Detection Methods

- **Alkaline Phosphatase**

Alkaline phosphatase-conjugated second antibody is reacted with a bromochloroindolyl phosphate - nitro blue tetrazolium substrate to give a dark purple precipitate. An important advantage over the horseradish peroxidase reaction is that the colored alkaline phosphatase product is stable and will not fade.

Follow Immunodetection Protocol (Section C) through step 6b, substituting an alkaline phosphatase-conjugated second antibody in step 5b.

7. Development (from Harlow and Lane, 1988)

    a. Prepare developing reagents

    Reagent Solutions
        (can be prepared in advance and stored at 4°C for >1 year)
        0.1 M Tris-HCl (pH 9.5), 0.1 M NaCl, 5 mM MgCl$_2$
            (Alkaline Phosphatase Buffer)
        50 mg/ml 5-bromo-4-chloro-3-indolyl phosphate in 100%
            dimethylformamide (BCIP Solution)
        50 mg/ml *p*-nitro blue tetrazolium chloride in 70%
            dimethylformamide (NBT Solution)

    b. Wash nitrocellulose for 5 minutes in Alkaline Phosphate Buffer.

    c. Prepare Developing Reagent (use within 1 hour):

        66 µl NBT Solution
        10 ml Alkaline Phosphatase Buffer
            *** Mix well at this point ***
        Add 33 µl BCIP Solution

    d. Pour off Alkaline Phosphatase Buffer and add 10 ml Developing Reagent to nitrocellulose.

    e. Incubate at room temperature or at 37°C to speed reaction.

    f. Reaction is mostly complete within 30 minutes, but it can be permitted to continue overnight to increase signal, and can be stopped by rinsing filter with 20 mM EDTA in TBS.

8

- **Immunogold**

Colloidal gold conjugates give a stable signal after development and are very easy and rapid to use. Unfortunately, high sensitivity requires a more complicated silver-enhancement step following initial development.

Follow Immunodetection Protocol through step 4b.

5. Development (from Hsu, 1984)

   a. Prepare gold-conjugated second antibody by diluting to A_{525}=0.5 in 0.05% Tween 20 in Tris-Buffered Saline (TBS).
   b. Remove TBS from nitrocellulose membrane and add second antibody solution.
   c. Incubate for 30 - 60 minutes.
   d. Rinse membrane with water after development.

6. Silver Enhancement of Immunogold Signal (Brada and Roth, 1984)

   a. After development (step 5c), rinse membrane twice for 5 minutes in TBS.
   b. Rinse membrane for 1 minute in distilled water.
   c. Dip nitrocellulose membrane into developer solution for 1 - 2 minutes (see Danscher, 1981).
   d. Wash membrane in tap water for 1 minute.
   e. Incubate membrane for 10 minutes in photographic fixer.

Color produced with gold stain is red, subsequent silver enhancement produces a black color (Brada and Roth, 1984).
A brief centrifugation just before use (250 - 4000 x g) removes aggregated gold particles. 5 nm gold particles give more sharply focused bands than 15 nm particles (Surek and Latzko, 1984).

- **^{125}I Methods**

^{125}I detection systems are in disfavor due to radioisotope handling and disposal problems, short half-lives, long wash steps to remove background, long development times and high cost (Bio-Rad Bulletin 1310, 1987). Protein A is less desirable than a 2nd antibody, since it does not recognize IgGs of all species or IgG subtypes. In addition, its binding is not polyvalent (Allen et al., 1984).

Follow Immunodetection Protocol through step 6b, substituting ^{125}I-labeled second antibody or ^{125}I-Protein A. **All solutions used after addition of the radiolabeled protein must be handled as radioactive waste.** 5 - 10 μCi (Kaufmann et al., 1987) or 2 - 5 x 10^6 cpm (Renart and Sandoval, 1987) of ^{125}I-labeled probe have been used.

8

7. Detection (from Kaufmann et al., 1987)

    a. Dry nitrocellulose thoroughly. Radioactive ink may be used to mark corners or molecular weight markers on the membrane.

    b. Wrap the membrane in plastic wrap.

    b. Expose to Kodak X-Omat XAR-5 or XRP-5 film with an intensifying screen at -70°C.

    c. Develop film according to instructions.

E. Staining for Total Protein

Staining for total protein on a membrane may be useful for monitoring the efficiency of transfer or for identifying an immunochemically detected band. For example, it is better to monitor the transfer efficiency in one or two lanes before probing the rest of the lanes with expensive antibody reagents. Three widely used methods for staining of nitrocellulose membranes are Amido Black, India ink, and Ponceau S staining. Nonspecific binding of anionic dyes makes these three methods less satisfactory for nylon membranes, but a biotin-avidin-horseradish peroxidase stain should provide satisfactory results (Bio-Rad Bulletin 1310, 1987). Colloidal gold staining provides the highest sensitivity.

Detection Levels:
   Amido Black - 30 ng per band (Bers and Garfin, 1985)
   India ink - 6 ng per band (Glenney, 1986)
   Enhanced Colloidal Gold - 400 pg per band (Bio-Rad Bulletin
      1310, 1987)
   Biotin - 10 - 50 ng per band (Bio-Rad Bulletin 1310, 1987)

- **Amido Black Staining** of nitrocellulose membranes:
  1. Stain 1 minute in 0.1% Amido Black 10B/25% isopropanol/10% acetic acid.
  2. Destain for 30 minutes in 25% isopropanol/10% acetic acid.
  3. Wash filters in TBS or water before drying to reduce shrinkage (Gershoni and Palade, 1982).

- **India Ink Staining** of nitrocellulose membranes:
  1. Wash membrane twice for 5 minutes in 0.5% Tween 20/TBS.
  2. Stain for at least 2 hours in 1 μl india ink/ml distilled water.
  3. Destain by rinsing several times for 5 minutes with distilled water. (Glenney, 1986).

- **Ponceau S Staining** of nitrocellulose membranes:
  1. Stain for 5 - 10 minutes in 0.2% Ponceau S/3% TCA/3% sulfosalicylic acid
  2. Wash away stain with TBS.
  3. Ponceau S staining can be followed by immunodecoration (Harlow and Lane, 1988).

• **Comments on Total Protein Staining**

1. Staining with anionic dyes in methanolic solutions will cause some shrinking of the nitrocellulose, so India ink staining may be preferred if exact replicas of blots are required.

2. No detectable quenching of radiolabel has been observed due to India ink staining (Glenney, 1986).

3. Certain protein bands are differentially stained with different staining methods. For example, acidic proteins stain only at pH 3.5 with colloidal gold (Rohringer and Holden, 1985).

4. Ponceau S staining is not very sensitive and the red color is difficult to photograph; however, the stain may be washed away and immunodetection may follow the staining.

8

F. Erasing Immunoblots

Erasing a blot involves removal of the primary and secondary antibodies. This permits repetitive use of a single blot. It should be noted that some epitopes will be damaged by the erasure treatment.

Erasing Buffer, 100 ml
    6.25 ml 1 M Tris-HCl (pH 6.8) 62.5 mM
    20 ml 10% SDS 2%
    0.7 ml 2-Mercaptoethanol 100 mM
    73 ml H_2O

- **Steps**

  1. After development of 2nd antibody reaction or prior to drying for autoradiography, incubate nylon or nitrocellulose membrane in 5% powdered nonfat milk/TBS for 10 minutes at room temperature.

  2. Dry on absorbent paper at room temperature. After 5 minutes, move to a fresh piece of absorbent paper to prevent membrane from sticking.

  3. To erase, incubate the dried membrane for 30 minutes at 70°C in Erasing Buffer.

  4. Following two 10 minute washes in TBS, nitrocellulose membranes should be incubated with 5% nonfat dry milk/TBS for 6 hours and nylon membranes should be recoated with 10% nonfat dry milk for 6 - 8 hours prior to reprobing with a new first antibody (Kaufmann et al., 1987).

- **Alternative erasing conditions**: see Renart and Sandoval (1984), Earnshaw and Rothfield (1985), and Surek and Latzko (1984).

III. Discussion

A. Membrane Storage

- To reduce the tendency of nitrocellulose membranes to stick to absorbent paper when drying, first incubate the blots for 10 minutes in nonfat dry milk/TBS, then dry 5 minutes and move to fresh absorbent paper (Kaufmann et al., 1987).

- Nitrocellulose paper can be stored for as long as a year following transfer before probing (Gershoni and Palade, 1982).

B. Transfer Anomalies

- Sometimes, proteins do not electroelute efficiently because they are fortuitously at their isoelectric point. In this case, other buffer conditions should be used (Gershoni and Palade, 1983). Strongly basic proteins such as histones, lysozymes or cytochromes may transfer poorly (Szewczyk and Kozloff, 1985).

- For isoelectric focusing gels, SDS-urea gels or nondenaturing gels containing basic proteins, transfer in 0.7% acetic acid with **gel cassette in reverse orientation** (that is, transferring toward the cathode).

- Exceeding the membrane binding capacity may reduce the signal in subsequent detection steps (Gershoni and Palade, 1983).

C. References for Other Uses of Immunoblots

- **Epitope Mapping:** Glenney et al., 1983; Mendelson et al., 1984.

- **Structural Domain Analysis:** Russel et al., 1984; Yurchenco et al., 1982.

- **Dot Blot:** May be applied to analysis of column fractions, sucrose gradients or pulse-chase experiments (Hawkes et al., 1982; Bosman et al., 1983; Talbot et al., 1984).

- **Functional Assay:**

    Although most functional tests for electroblotted proteins currently involve ligand binding assays (see section below), improved protein renaturation techniques should allow the establishment of other specific enzymatic assays. Enzymatic assays have been described for proteins in polyacrylamide gels, including dehydrogenases, phosphatases, esterases, oxidases, and peptidases, and these are referenced in Chapter 6, section III.B.

    Conditions for preserving or restoring enzyme activity during electroblotting may involve sample treatment prior to electrophoresis, gel incubation prior to electrotransfer, or incubation of the membrane after protein transfer. Protein sample treatments before and during gel electrophoresis include eliminating the use of sulfhydryl reagents (Islan et al., 1983; Daniel et al., 1983), EDTA (Gershoni et al., 1983), or sample heating (Daniel et al., 1983). Polyacrylamide gel treatment may involve washing the gel in a buffered solution to remove SDS (Bowen et al., 1980; Wolff et al., 1985). Nitrocellulose membranes have been incubated in solutions containing low amounts of detergent to allow protein renaturation (Haeuptle et al., 1983).

- **Ligand Binding:**

    A wide variety of ligands have been used as probes for proteins after immunoblotting, including:

    DNA (Bowen et al., 1980; Hoch, 1982; Patel and Cook, 1983)

    RNA (Bowen et al., 1980; Rozier and Mache, 1984)

    Lectins (Hawkes, 1982)

    Hormones (Haeuptle et al., 1983)

    Toxins (Gershoni and Palade, 1983)

    Viruses (Co et al, 1985)

    Heparin (Cardin et al., 1984)

    Calmodulin (Gershoni and Palade, 1983; Flanagan and Yost, 1984)

Histones (Bowen et al., 1980)

Whole Cells (Hayman et al., 1982)

GTP (McGrath et al., 1984)

Calcium (Fong et al., 1988)

Zinc (Serrano et al., 1988)

- **Purifying Antibody From an Immunoblot:** Smith and Fisher, 1984; Olmsted, 1981.

- **Cutting Protein Bands From a Nitrocellulose Membrane for Antibody Production:** Knudson, 1985; Harlow and Lane, 1988, p. 498.

- **Protein Identification: Amino Acid Analysis and Protein Sequencing**

    Minute amounts of protein (as little as 10 pmol) may be analyzed for amino acid composition (Tous et al., 1989) or sequence (Matsudaira, 1987) after electroblotting on poly(vinylidene difluoride) (PVDF) membranes. Transferred protein or peptide bands are stained with Coomassie blue and then excised from the membrane.

8

IV. Suppliers

Alkaline Phosphatase Conjugates: Bio-Rad; Calbiochem; Sigma

Colloidal Gold and biotin total protein staining kits: Bio-Rad

Colloidal Gold Conjugates: Bio-Rad; Sigma

Electroblotting Apparatus: Bio-Rad Mini Trans-Blot Electrophoretic Transfer Cell; Hoefer Model TE 22 Mighty Small Transfer Unit; Schleicher and Schuell Mini Transfer System

Filter Paper: Bio-Rad; Schleicher and Schuell; Whatman

Heat-Sealable Plastic Pouches and Sealers
- **Kapak:** Fisher, Thomas, VWR
- **Seal-A-Meal:** Local retailers

Horseradish Peroxidase Conjugates: Bio-Rad; Calbiochem; Sigma

India Ink: Pelikan fount india drawing ink, Pelikan AG; Speedball drawing ink, dense black india, No. 3211, Hunt Mfg. Co.

Nitrocellulose: Bio-Rad; Millipore; Schleicher and Schuell

Nylon: Bio-Rad (Zeta-probe); CUNO (Zetabind)

Ponceau S (3-hydroxy-4-[2-sulfo-4-(sulfo-phenylazo)phenylazo]-2,7-naphthalene disulfonic acid): Sigma

Power supply with a capacity of 200 V, 0.6 A: Bio-Rad, Pharmacia

Rocker: Hoefer Red Rocker

V. References

Allen, R.C., C.A. Saravis and H.R. Maurer. 1984. Gel Electrophoresis and Isoelectric Focusing of Proteins. Selected Techniques. Walter de Gruyter, Berlin. pp. 221-230.

Bers, G. and D. Garfin. 1985. BioTechniques 3: 276-288. Protein and Nucleic Acid Blotting and Immunobiochemical Detection.

Bio-Rad Bulletin 1310. 1987. Western Blotting Detection Systems: How do you choose?

Bio-Rad Immun-Blot (GAR-HRP) Assay Kit Bulletin.

Bio-Rad Mini Trans-Blot Electrophoretic Transfer Cell Instruction Manual.

Blake, M.S., K.H. Johnston, G.J. Russel-Jones and E.C. Gotschlich. 1984. Anal. Biochem. 136: 175-179. A Rapid, Sensitive Method for Detection of Alkaline Phosphatase-Conjugated Anti-Antibody on Western Blots.

Bosman, F.T., G. Cramer-Knijnenburg and J.v.B. Henegouw. 1983. Histochemistry 77: 185-194. Efficiency and Sensitivity of Indirect Immunoperoxidase Methods.

Bowen, B., J. Steinberg, U.K. Laemmli and H. Weintraub. 1980. Nuc. Acids Res. 8: 1-20. The Detection of DNA-Binding Proteins by Protein Blotting.

Brada, D. and J. Roth. 1984. Anal. Biochem. 142: 79-83. "Golden Blot" - Detection of Polyclonal and Monoclonal Antibodies Bound to Antigens on Nitrocellulose by Protein A - Gold Complexes.

Burnette, W.N. 1981. Anal. Biochem. 112: 195-203. "Western Blotting": Electrophoretic Transfer of Proteins From Sodium Dodecyl Sulfate - Polyacrylamide Gels to Unmodified Nitrocellulose and Radiographic Detection With Antibody and Radioiodinated Protein A.

Cardin, A.D., K.R. Witt and R.L. Jackson. 1984. Anal. Biochem. 137: 368-373. Visualization of Heparin-Binding Proteins by Ligand Blotting with ^{125}I-Heparin.

Co, M.S., G.N. Gaulton, B.N. Fields and M.I. Greene. 1985. Proc. Natl. Acad. Sci. USA 82: 1494-1498. Isolation and Biochemical Characterization of the Mammalian Reovirus Type 3 Cell-Surface Receptor.

Daniel, T.O., W.J. Schneider, J.L. Goldstein and M.S. Brown. 1983. J. Biol. Chem. 258: 4606-4611. Visualization of Lipoprotein Receptors by Ligand Blotting.

Danscher, G. 1981. Histochemistry 71: 81-88. Localization of Gold in Biological Tissue.

Earnshaw, W.C. and N. Rothfield. 1985. Chromosoma 91: 313-321. Identification of a Family of Human Centromere Proteins Using Autoimmune Sera from Patients with Scleroderma.

Flanagan, S.D. and B. Yost. 1984. Anal. Biochem. 140: 510-519. Calmodulin-Binding Proteins: Visualization by ^{125}I-Calmodulin Overlay on Blots Quenched with Tween 20 or Bovine Serum Albumin and Poly(ethylene oxide).

Fong, K.C., J.A. Babitch and F.A. Anthony. 1988. Biochim. Biophys. Acta 952: 13-19. Calcium Binding to Tubulin.

Gershoni, J.M., E. Hawrot and T.L. Lentz. 1983. Proc. Natl. Acad. Sci. USA 80: 4973-4977. Binding of Alpha-Bungarotoxin to Isolated Alpha Subunit of the Acetylcholine Receptor of *Torpedo californica*: Quantitative Analysis With Protein Blots.

Gershoni, J.M. and G.E. Palade. 1982. Anal. Biochem. 124: 396-405. Electrophoretic Transfer of Proteins From Sodium Dodecyl Sulfate - Polyacrylamide Gels to a Positively Charged Membrane Filter.

Gershoni, J.M. and G.E. Palade. 1983. Anal. Biochem. 131: 1-15. Protein Blotting: Principles and Application.

Glenney, J. 1986. Anal. Biochem. 156: 315-319. Antibody Probing of Western Blots Which Have Been Stained with India Ink.

Glenney, Jr., J.R., P. Glenney and K. Weber. 1983. J. Mol. Biol. 167: 275-293. Mapping the Fodrin Molecule with Monoclonal Antibodies.

Haeuptle, M.-T., M.L. Aubert, J. Djiane and J.-P. Kraehenbuhl. 1983. J. Biol. Chem. 258: 305-314. Binding Sites for Lactogenic and Somatogenic Hormones from Rabbit Mammary Gland and Liver.

Harlow, E. and D. Lane. 1988. Antibodies: A Laboratory Manual. 726 pages. Cold Spring Harbor Laboratory, Cold Spring Harbor, New York.

Hawkes, R. 1982. Anal. Biochem. 123: 143-146. Identification of Concanavalin A-Binding Proteins after Sodium Dodecyl Sulfate - Gel Electrophoresis and Protein Blotting.

Hawkes, R., E. Niday and J. Gordon. 1982. Anal. Biochem. 119: 142-147. A Dot-Immunobinding Assay for Monoclonal and Other Antibodies.

Hayman, E.G., E. Engvall, E. A'Hearn, D. Barnes, M. Pierschbacher and E. Ruoslahti. 1982. J. Cell Biol. 95: 20-23. Cell Attachment on Replicas of SDS Polyacrylamide Gels Reveals Two Adhesive Plasma Proteins.

Hoch, S.D. 1982. Biochem. Biophys. Res. Comm. 106: 1353-1358. DNA-Binding Domains of Fibronectin Probed Using Western Blots.

Hsu, Y.-H. 1984. Anal. Biochem. 142: 221-225. Immunogold for Detection of Antigen on Nitrocellulose Paper.

Islan, M.N., R. Briones-Urbina, G. Bako and N.R. Farid. 1983. Endocrin. 113: 436-438. Both TSH and Thyroid-stimulating Antibody of Graves' Disease Bind to an M_r 197,000 Holoreceptor.

Johnson, D.A., J.W. Gautsch, J.R. Sportsman and J.H. Elder. 1984. Gene Analysis Techniques 1: 3-8. Improved Technique Utilizing Nonfat Dry Milk for Analysis of Proteins and Nucleic Acids Transferred to Nitrocellulose.

Kaufmann, S.H., C.M. Ewing and J.H. Shaper. 1987. Anal. Biochem. 161: 89-95. The Erasable Western Blot.

Knudson, K.A. 1985. Anal. Biochem. 147: 285-288. Proteins Transferred to Nitrocellulose for Use as Antigens.

McGrath, J.P., D.J. Capon, D.V. Goeddel and A.D. Levinson. 1984. Nature 310: 644-649. Comparative Biochemical Properties of Normal and Activated Human *ras* p21 Protein.

Matsudaira, P. 1987. J. Biol. Chem. 262: 10035-10038. Sequence from Picomole Quantities of Proteins Electroblotted onto Polyvinylidene Difluoride Membranes.

Mendelson, E., B.J. Smith and M. Bustin. 1984. Biochem. 23: 3466-3471. Mapping the Binding of Monoclonal Antibodies to Histone H5.

Olmsted, J.B. 1981. J. Biol. Chem. 256: 11955-11957. Affinity Purification of Antibodies from Diazotized Paper Blots of Heterogeneous Protein Samples.

Otter, T., S.M. King and G.B. Witman. 1987. Anal. Biochem. 162: 370-377. A Two-Step Procedure for Efficient Electrotransfer of Both High-Molecular-Weight (>400,000) and Low-Molecular-Weight (<20,000) Proteins.

Patel, S.B. and P.R. Cook. 1983. EMBO J. 2: 137-142. The DNA-Protein Cross: A Method for Detecting Specific DNA-Protein Complexes in Crude Mixtures.

Peluso, R.W. and G.H. Rosenberg. 1987. Anal. Biochem. 162: 389-398. Quantitative Electrotransfer of Proteins from Sodium Dodecyl Sulfate-Polyacrylamide Gels onto Positively Charged Nylon Membranes.

Renart, J. and I.V. Sandoval. 1984. Meth. Enzymol. 104: 455-459. Western Blots.

Rohringer, R. and D.W. Holden. 1985. Anal. Biochem. 144: 118-127. Protein Blotting: Detection of Proteins with Colloidal Gold, and of Glycoproteins and Lectins with Biotin-Conjugated and Enzyme Probes.

8

Rozier, C. and R. Mache. 1984. Nuc. Acids Res. 12: 7293-7304. Binding of 16S rRNA to Chloroplast 30S Ribosomal Proteins Blotted on Nitrocellulose.

Russel, D.W., W.J. Schneider, T. Yamamoto, K.J. Luskey, M.S. Brown and J.L. Goldstein. 1984. Cell 37: 577-585. Domain Map of the LDL Receptor: Sequence Homology with the Epidermal Growth Factor Precursor.

Serrano, L., J.E. Dominguez and J. Avila. 1988. Anal. Biochem. 172: 210-218. Identification of Zinc-Binding Sites of Proteins: Zinc Binds to the Amino-Terminal Region of Tubulin.

Smith, D.E. and P.A. Fisher. 1984. J. Cell Biol. 99: 20-28. Identification, Developmental Regulation, and Response to Heat Shock of Two Antigenically Related Forms of a Major Nuclear Envelope Protein in Drosophila Embryos: Application of an Improved Method for Affinity Purification of Antibodies Using Polypeptides Immobilized on Nitrocellulose Blots..

Surek, B. and E. Latzko. 1984. Biol. Biochem. Res. Comm. 121: 284-289. Visualization of Antigenic Proteins Blotted onto Nitrocellulose Using the Immuno-Gold-Staining (IGS)-Method.

Szewczyk, B. and L.M. Kozloff. 1985. Anal. Biochem. 150: 403-407. A Method for the Efficient Blotting of Strongly Basic Proteins from Sodium Dodecyl Sulfate-Polyacrylamide Gels to Nitrocellulose

Talbot, P.J., R.L. Knobler and M.J. Buchmeier. 1984. J. Immunol. Methods 73: 177-188. Western and Dot Immunoblotting Analysis of Viral Antigens and Antibodies: Application to Murine Hepatitis Virus.

Tous, G.I., J.L. Fausnaugh, O. Akinyosoye, H. Lackland, P. Winter-Cash, F.J. Vitorica and S. Stein. 1989. Anal. Biochem. 179: 50-55. Amino Acid Analysis on Polyvinylidene Difluoride Membranes.

Towbin, H. and J. Gordon. 1984. J. Immunol. Meth. 72: 313-340. Immunoblotting and Dot Immunoblotting - Current Status and Outlook.

Towbin, H., T. Staehelin and J. Gordon. 1979. Proc. Nat. Acad. Sci. 76: 4350-4354. Electrophoretic Transfer of Proteins from Polyacrylamide Gels to Nitrocellulose Sheets: Procedure and Some Applications.

Van Eldik, L.J. and S.R. Wolchok. 1984. Biochem. Biophys. Res. Comm. 124: 752-759. Conditions for Reproducible Detection of Calmodulin and S100B in Immunoblots.

Wolff, P., R. Gilz, J. Schumacher and D. Riesner. 1985. Nucl. Acids Res. 13: 355-367. Complexes of Viroids with Histones and Other Proteins.

Yurchenco, P.D., D.W. Speicher, J.S. Morrow, W.F. Knowles and V.T. Marchesi. 1982. J. Biol. Chem. 257: 9102-9107. Monoclonal Antibodies as Probes of Domain Structure of the Spectrin a Subunit.

8

Purification and Crystallization of Proteins

The second portion of *Protein Methods* focuses on protein purification and crystallization. Because of the nature of these techniques, it is not possible to develop universally applicable protocols. Instead, general approaches must be tailored by the investigator according to the needs of the specific protein of interest. For example, chromatographic methods are often used in trial experiments to probe the basic physical and chemical characteristics of proteins. Subsequent experiments then make use of this information to design the most efficient purification protocol. Similarly, initial crystallization experiments screen a wide range of conditions to find those in which the solubility of the protein of interest is minimized. The most promising are then used as the basis for a more narrow search of conditions in an optimization screen. Accordingly, in describing these techniques, we have reduced the specificity of the experimental descriptions while keeping a general, step-by-step framework for the investigator to use in formulating his or her own protocol.

The following chapters outline the most general procedures for three types of chromatography (ion exchange, gel filtration, and affinity) as well as the hanging drop method of crystallization. Protein purification and crystallization are often considered to be equal parts of science and art, since a number of variables can tip the balance between an elegant separation or the growth of large crystals on one side and a spoiled, contaminated protein soup on the other. Often, the investigator relies more on intuition about the behavior of the protein than on adherence to strict rules of protocol. Insufficient attention to the requirements of the protein (pH and temperature optima, buffer composition, and the presence or absence of cofactors and salts) or poor execution of technique can cost precious time and protein.

When embarking on a protein purification, several general rules are important to keep in mind. First, a suitable assay for protein activity should be available and should be utilized at each step of the purification to avoid wasting a significant effort on purifying an inactive protein. The assay should require minimal amounts of protein and should not take a long time to perform. Suitable assays could measure enzyme activity, ligand binding, or some other unique property of the protein of interest. Second, since a protein can rarely be purified to homogeneity in a single step, a researcher must determine not only which procedures to use but also in what order these procedures should be carried out. The specific behavior and characteristics of the protein of interest should always guide these decisions. Finally, the purification steps should exploit complementary properties of the protein such as charge, size or hydrophobicity to be most effective. Frequently, protein purifications employ ion exchange chromatography as a first step, since it is nearly always effective at providing some degree of purification in a crude extract and it can accommodate the large volumes that are often obtained for crude extracts. This step would be followed by gel filtration chromatography, which rarely inactivates enzymes because it does not involve binding to the chromatography matrix. The last purification step might be an affinity column, which offers the highest degree of specificity but is often too expensive to be used on a large scale with cruder preparations.

Chromatography experiments such as those described in the following chapters are analyzed by comparing the total protein eluted in each fraction with a measure of the activity of the protein of interest. The **specific activity** of the protein is the amount of functional activity, usually measured in units of activity, divided by the total amount of protein in milligrams. The degree of purification of the protein at each stage of the purification is generally indicated by an increase in the specific activity. However, since inactivation of the protein will cause the specific activity to decrease, it is usually advisable to check the purity of the protein by an independent means, such as SDS gel electrophoresis (Chapter 5). If further purification is required, the fractions containing the protein of interest are then pooled in preparation for a subsequent step.

Chapter 9

Ion Exchange Chromatography

I. Protein Purification
    A. Introduction
    B. Methods

II. Concentrating a Protein Solution

III. Batch Chromatography

IV. Suppliers

V. References

9

I. Protein Purification

A. Introduction

Ion exchange chromatography is the most commonly used chromatographic method for protein purification. Its popularity stems from the possibility of high resolution protein separation, the relative ease of use and reproducibility, and the availability and low cost of materials. The ion exchange principle permits the protein to bind even when a large buffer volume is applied, making this method especially useful for an initial purification step from a crude extract. For separations in which speed is important (such as removal of proteases in a crude extract or purification of proteins with limited activity lifetimes), ion exchange chromatography offers many support materials with rapid flow rates. These properties make ion exchange chromatography a valuable step in most protein purifications and an excellent starting point for a purification protocol.

Ion exchange chromatography requires that a protein contain a net ion charge under experimental conditions (for definition of net charge, see Chapter 7, Section I). As a result, the protein will displace a low molecular weight ion from an ion exchange matrix (an 'exchange of ions', hence the term ion exchange) and become bound. A change in experimental conditions (such as an increase in the counter-ion concentration or a decrease in the protein net ion charge) will cause another exchange of ions, this time releasing the protein from the ion exchange matrix in favor of the counter-ion. This process of successive exchanges of ions allows the separation of proteins with different charge properties.

When the isoelectric point of the protein is known, an initial strategy for separation can be easily developed (see Section 2 below). On the other hand, when little is known about the protein, a few introductory ion exchange experiments can reveal much about the biophysical properties of a protein and can guide the subsequent purification.

1. Ion Exchange Principles

Why is knowledge of the isoelectric point important in ion exchange chromatography?

Like isoelectric focusing (Chapter 7), separation and purification of proteins using ion exchange chromatography is based primarily on differences in the ionic properties of surface amino acids. Exposed arginine, histidine, and lysine residues are generally positively charged, and aspartic acid and glutamic acid residues possess a negative charge at neutral pH. Thus, at a given pH, a protein will possess an overall **net charge**. At a lower pH, the net charge will be more positive, and at a higher pH, the net charge will be more negative (Fig. 9.1). The pH at which the positive charges equal the negative charges (in other words, the net charge of the protein is zero) defines that protein's **isoelectric point** (pI). For ion exchange chromatography, a good rule to follow when separating a protein whose isoelectric point is known is to select a working pH which is 1 unit away from the pI of the protein. At this pH, the protein will possess a high enough net charge to bind well to the ion exchange column without being so highly charged that harsh elution conditions are required (such as extremely high ionic strength or significant change in pH).

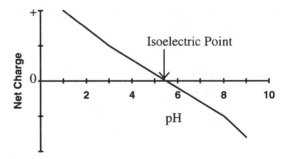

Figure 9.1. Effect of pH on protein net charge.

How does an ion exchanger bind protein?

Assuming that the protein possesses a net charge, an ion exchanger needs an opposite charge in order to bind that protein. Ion exchangers are typically composed of a charged (ion exchange) group attached to an insoluble matrix. A positively charged group, such as DEAE (diethylamino ethyl), defines the matrix as an **anion exchange matrix** whereas a negatively charged group, such as CM (carboxymethyl) makes a **cation exchange matrix**.

DEAE $-CH_2CH_2N^+H(CH_2CH_3)_2$

CM $-CH_2COO^-$

Thus, if a protein has a negative net charge at a given pH, an anion exchange matrix should be used for its purification. The matrix material is usually a polymer such as agarose, cellulose, or acrylamide in the form of tiny beads cross-linked to varying degrees. Differences in matrix materials and ion exchangers are discussed in Section B.2 below.

What is the procedure for protein purification?

Separation of proteins by ion exchange chromatography requires differential binding of the proteins to the ion exchange matrix. After proteins are applied to an ion exchange column, those proteins which have no affinity for the matrix are removed during a wash of the column. Then, the adsorbed proteins are removed in an elution step. Since different proteins generally bind to an ion exchange matrix with differing affinities, it is usually possible to separate them by gradually increasing the salt concentration during the elution step. For example, proteins that are negatively charged near neutral pH would stick to a positively-charged anion exchange matrix. The act of binding would displace a counter-ion (such as Cl⁻ with a loading buffer containing 0.1 M chloride) from the ion exchanger. If a higher concentration of the counter-ion were applied (for example, 0.3 M Cl⁻), a protein with a low net charge might be displaced. To remove another protein with a greater net charge, 0.5 M chloride may be required. By progressively raising the counter-ion concentration, a clean separation of the two proteins can be achieved. A useful alternative strategy for purification is to select conditions so that the protein of interest does not bind to the ion exchange matrix while contaminating substances are adsorbed.

What protein elution strategies are available?

A protein binds reversibly to an ion exchange column by electrostatic forces. The most frequently used strategy for disrupting the electrostatic force between protein and ion exchanger is to raise the counter-ion (salt) concentration. Counter-ions (such as Na^+ or Cl⁻) are low molecular weight ions which, at low concentrations, can be dissociated from an ion exchanger by a protein, and which, at elevated concentrations, can effectively compete with that protein for binding to the ion exchanger. Besides being effective and inexpensive, a salt elution is simple to perform and is easily reproducible. Two methods exist for performing a salt elution. In a **step elution**, the salt concentration is increased in distinct steps. A **gradient elution** utilizes a gradient maker to establish a smooth (continuous) increase in salt concentration.

How do different counter-ions affect the elution?

For a given ion exchanger (such as DEAE), different counter-ions (for example, Cl⁻, PO_4^{3-}, HCO_3^-) may be used for eluting an adsorbed protein. Counter-ions possess different affinities for an ion exchanger, and the relative affinities are summarized in an activity series (see Table 9.1). Thus, a protein which fails to elute from an ion exchange column with one counter-ion may be displaced when a stronger counter-ion is used.

Table 9.1
Counter-ion Activity Series

For anion exchange:

$Ag^+ >$ (binds more tightly than) $Cs^+ > K^+ > NH_4^+ > Na^+ >$ $H^+ > Li^+$

For cation exchange:

$I^- > NO_3^- > PO_4^{3-} > CN^- > HSO_3^- > Cl^- > HCO_3^- > HCOO^-$ $> CH_3COO^- > OH^- > F^-$

Are other elution strategies useful?

An alternative protein elution method is to change the buffer pH so as to lower the protein binding affinity. Lowering the pH when using an anion exchanger weakens the affinity of a bound protein (as does raising the pH with a cation exchanger). The weaker affinity is due to a reduction in the net charge of the protein (see Fig. 9.1). As the buffer pH approaches the pI of a given protein, the protein becomes less tightly bound to the matrix and ultimately is eluted from the column. Indeed, **chromatofocusing** is a specialized form of ion exchange chromatography predicated on separating proteins on the basis of differences in isoelectric points. Giri (1990), Li and Hutchens (1992), and Roe (1989) provide useful descriptions and protocols for chromatofocusing. In the absence of chromatofocusing resins and buffers, pH changes can cause temporary changes in ionic strength in the ion exchange column; consequently, changing the buffer pH is a rarely used elution method for ion exchange chromatography.

What buffer should I use?

The ion exchange buffer must be carefully chosen, since a poorly buffered solution will allow pH fluctuations which can affect the protein binding affinity (Fig. 9.1). As explained in Section 2 below, the experimental buffer should be chosen to provide adequate buffering capacity one pH unit away from a protein's isoelectric point. Remember that the buffering capacity is highest if the experimental pH is within one unit of the buffer's pKa, and preferably within 0.4 units. To further enhance the buffering capacity, the minimal ionic strength of the counter-ion should be at least 10 - 20 mM. Since a chromatography buffer contains buffer ions as well as counter-ions, the charged form of the buffer should carry the same charge as the ion exchanger (for example, $HTris^+$ with a DEAE exchanger); otherwise, the buffer ion can bind to the ion exchanger, reducing the buffering capacity and possibly causing local pH fluctuations. Table 9.2 provides some buffer suggestions for ion exchange chromatography.

Table 9.2
Buffers for Ion Exchange Chromatography

pKa	pH Range	Buffer
		For Cation Exchange
3.8	3.4 - 4.2	Lactic Acid
4.8	4.4 - 5.2	Acetic Acid
6.2	5.8 - 6.6	MES (2-(N-Morpholino)ethanesulfonic acid)
6.6	6.2 - 7.0	ADA (N-2-Acetamidoiminodiacetic acid)
7.2	6.8 - 7.6	MOPS (3-(N-Morpholino)propanesulfonic acid)
		For Anion Exchange
6.0	5.6 - 6.4	Histidine
7.8	7.4 - 8.2	Triethanolamine
8.1	7.7 - 8.5	Tris
8.8	8.4 - 9.2	Diethanolamine
9.5	9.1 - 9.9	Ethanolamine

See also Scopes (1994) and Roe (1989)

2. A General Guide

When setting out to purify a protein by ion exchange chromatography, the most important decisions are 1) selecting the best ion exchange matrix and buffer for the separation and 2) determining how to elute the protein of interest. In this section, we provide some guidelines for making these decisions.

a. Choosing the Right Matrix and Buffer

Knowing a protein's isoelectric point simplifies the process of devising an effective ion exchange protocol. Since the pI of a protein may not always be known, this section is divided into two parts.

1) When the isoelectric point is known

The ion exchange matrix is selected according to a protein's net charge under chromatographic conditions. If the protein is negatively charged, it will bind to a positively charged anion exchange matrix, and a protein which is positively charged will bind to a cation exchange matrix. For example, a protein with a pI of 6.9 will be negatively charged above that pH and positively charged below. Table 9.3 provides information about specific matrices.

The isoelectric point indicates how strongly charged a protein will be at a given pH (see Fig. 9.1). In order to avoid conditions under which the protein of interest binds too tightly or too weakly to the column, a pH approximately one unit from the pI is the best choice for attempting a purification. Thus, for a protein with a pI of 6.9, buffers of pH 5.9 (and a cation exchange matrix such as CM Sepharose CL-6B) or 7.9 (with an anion exchanger such as DEAE Sepharose CL-6B) are recommended.

The second concern when choosing the best pH is protein stability. In our example, if the protein denatures below pH 6, a buffer of pH 7.9 is ideal (for example, Tris). Small scale experiments can confirm that experimental conditions do not impair protein activity. Consult Table 9.2 for specific buffers.

2) When the isoelectric point is not known

The most common way to determine an isoelectric point is with isoelectric focusing (Chapter 7). It should be noted that isoelectric points calculated from a primary amino acid sequence are frequently different from the observed pI by as much as an entire pH unit.

If the protein is impure and isoelectric focusing becomes impossible, a simple survey of pH conditions, presented below, is recommended by Pharmacia (1991). An activity assay will be required for determining the presence of the protein. The procedure below can aid in determining which pH is appropriate for an ion exchange purification.

Determining appropriate pH for ion exchange chromatography

a) Prepare 9 test tubes, each containing 1 ml of ion exchanger (e.g. DEAE Sepharose CL-6B).

b) Equilibrate the first tube with 0.5 M pH 5.0 buffer containing 10 mM NaCl by washing ten times with 10 ml of buffer. Equilibrate the second tube with pH 5.5 buffer, and so forth until the ninth test tube is equilibrated at pH 9.0.

c) Remove enough buffer so that approximately 1 ml of buffer covers the matrix.

d) Add 100 µl of protein solution to each tube.

e) Mix the tube and allow the matrix to settle for a few minutes.

f) Test the supernatant for the protein of interest.

The best conditions for ion exchange chromatography are those under which the protein is bound, but at a pH not far from one at which the protein dissociates. These conditions should permit adequate protein binding without requiring extreme efforts to elute the protein. In any event, a trial elution should be performed with a higher salt concentration, and protein activity should be tested. In the rare event that a protein does not bind under any of the pH conditions tested, the procedure should be repeated with a cation exchange matrix (such as CM Sepharose CL-6B). If the protein also does not bind to the cation exchanger, the best ion exchange purification for this protein may be to remove most other proteins by simply running the solution containing the protein straight through an anion and cation exchange matrix arranged in series (Scopes, 1994).

b. Choosing the Elution Conditions

Selecting the most appropriate counter-ion concentration will greatly simplify a protein purification. A trial experiment should be undertaken to find the salt concentration required to elute a given protein under the buffer conditions defined above. Such an experiment for the example protein from Section 2.a.i. follows:

1) Equilibrate 10 ml of DEAE-Sepharose CL-6B ion exchanger with 50 mM Tris buffer (pH 7.9) containing 10 mM NaCl.

2) Distribute 1 ml of ion exchanger into each of 9 test tubes.

3) Equilibrate the second tube with 0.1 M NaCl by washing ten times with 10 ml of 50 mM Tris buffer (pH 7.9) containing 0.1 M NaCl.

4) Equilibrate the third tube in the same fashion with pH 7.9 Tris buffer containing 0.2 M NaCl. Equilibrate the fourth tube with the same buffer containing 0.3 M NaCl, and continue on with the fifth tube (0.4 M NaCl), the sixth tube (0.5 M), the seventh (0.6 M), eighth (0.8 M), and ninth (1.0 M NaCl).

5) Remove buffer so that only about 1 ml of buffer covers the matrix.

6) Add 100 μl of protein solution to each tube.

7) Mix each tube and allow the matrix to settle for a few minutes.

8) Test the supernatants for the protein of interest.

The protein should appear in the supernatant at elevated salt concentrations, perhaps 0.4 M. If the protein already appears in the supernatant at low salt concentrations (below 0.1 M), changing the pH to increase the protein net charge may be useful (for example, increase the pH by 0.2 units to 8.1). Conversely, if the protein remains bound under high salt concentrations, a buffer pH closer to the isoelectric point may reduce the adsorption affinity or a different counter-ion with a higher binding affinity (such as PO_4^-) may dislodge the protein under more moderate conditions.

If the sample is subsequently loaded onto the column at a salt concentration just below that required for elution, the more weakly binding contaminating proteins will be removed during the wash step, simplifying the elution procedure. For example, if a protein elutes at 0.5 M NaCl, loading conditions of 0.3 M NaCl may work well. Remember, however, that loading and washing a column with a buffer which allows only limited adsorption of the protein can lead to slow, possibly undetectable protein leaching from the column. Often, a balance must be struck between a lengthy wash step which might result in a slow loss of the protein of interest and a brief wash which may leave more contaminants in the column.

A protein adsorbed on an ion exchange column can be released by step elution or gradient elution (see Section B.5). A step elution is generally preferred for convenience and reproducibility when high resolution is not essential or if the contaminants elute at a significantly different counter-ion concentration than the protein of interest. A gradient elution will usually provide better separation between closely spaced peaks, and can be defined by first running a step gradient or by following the example experiment outlined above. After the initial flow-through peak emerges from the column, the total volume of elution buffer for a gradient elution should be five to ten column volumes while three to ten column volumes should be used for each salt concentration in a step gradient. Usually all proteins will have eluted from a column at salt concentrations approaching 1 M. When resolution is not a major concern, batch elution can offer significant time savings (see Section III).

B. Methods

1. Equipment

a. A Basic Chromatography Setup

A standard ion exchange chromatography system contains the equipment shown in Fig. 9.2. A simpler setup may be appropriate for preliminary investigations and more complex systems can make routine or frequent separations more reproducible and less labor intensive, as described in the sections below. However, for most protein purifications, the basic system described in this section provides excellent results.

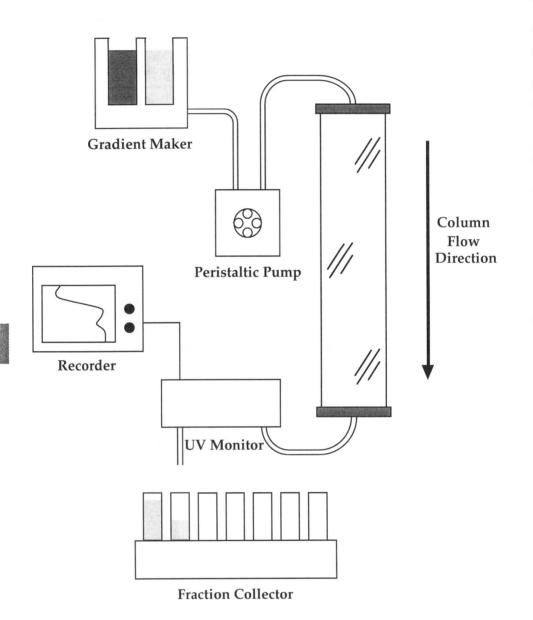

Figure 9.2. Ion exchange chromatography system.

1) **Column**. Generally a column consists of a glass or plastic cylinder with a thin porous plug at the bottom to support the ion exchange matrix. A minimal volume ("dead space") should exist between the porous support and the column outlet since sample remixing can occur in this space, reducing the resolution. Flow adapters for the top of the column are commercially available; flow adapters apply a sample or buffer solution evenly to the matrix surface and can be particularly useful during sample loading. Adapters are required when performing the chromatogram under positive pressure, for example when using high performance systems. For high resolution separations, the column height should be four to five times the diameter, whereas for faster, less rigorous separations, a height of one to two times the diameter is appropriate. Note that prepacked ion exchange columns are also available commercially.

2) **Tubing**. Pliable tubing is used to connect the column to the buffer reservoir and to the fraction collector. Materials such as silicone, Teflon, or Tygon tubing may be utilized.

3) **Fraction Collector**. The fraction collector automatically presents a new collection vial at the column outlet as a function of time or eluted volume (number of drops). When collecting fractions as a function of time, differences in column flow rate will result in fractions containing unequal volumes (this complication can be overcome with the use of a peristaltic pump). Fraction volumes can also vary when fractions are collected according to a specified number of drops, since the volume of a drop can vary depending on the protein concentration in the solution.

4) **UV Detector**. An ultraviolet wavelength detector can monitor light absorbance by proteins or peptides (at 280 or 230 nm for example) as they emerge from the column. Thus, a profile of the column eluent is directly provided. Alternatively, individual fractions can be assayed for UV absorbance with a spectrophotometer once they have been collected or another protein detection assay may be used (see Chapter 3).

5) **Recorder**. A chart recorder provides a tracing of the absorbance measured by the UV detector.

6) **Buffer Reservoir**. A reservoir holds the buffer solution until the solution enters the column. A beaker or Erlenmeyer flask is often used.

7) **Peristaltic Pump**. A pump can speed buffer flow through the column as well as maintain the flow at a constant rate. If a blockage occurs while a pump is running, however, a pressure buildup or burst tubing may result.

8) **Gradient Maker** (for gradient elutions). A gradient maker can be purchased commercially or constructed with two beakers and some tubing.

b. A Simple System

For preliminary screens or low-resolution, low-volume separations, a simple column can be made in a Pasteur pipet. Plug the end with some glass wool, and pack and run the column as described in the sections below. One milliliter fractions can even be collected by hand. Econo-Columns (from Bio-Rad) are also simple columns which can be attached to a simple chromatography setup.

c. Automated Systems

Automated chromatography systems which run under high pressure are extremely useful for workers who frequently use column chromatography for protein separation. These systems offer increased speed and generally greater reproducibility. Due to the specialized nature of these products, the reader is directed to the manufacturers for information about the design and operation of automated systems. The standard chromatography systems described in this text usually achieve equal levels of purification at a lower cost for initial protein purification efforts.

9

2. Selecting an Ion Exchange Matrix

If the protein to be purified is negatively charged under experimental conditions, an anion exchange matrix must be used and conversely, a positively charged protein will adsorb to a cation exchange matrix (see Section A.1). For most introductory experiments, an agarose-based ion exchanger such as DEAE- or CM-Sepharose is appropriate. Other ion exchangers are described in Table 9.3. Some of the properties which distinguish ion exchange matrices are briefly discussed below.

a. Type of exchanger

Ion exchangers, whether anion or cation exchangers, can be strong or weak exchangers. Strong ion exchangers remain fully ionized over most of the useful pH range while weak ion exchangers become less ionized as a function of pH. Thus, if chromatography conditions require pH extremes, a strong ion exchanger is required. Whereas most protein purifications in the past have relied on a weak exchanger, strong exchangers may provide equal levels of separation with greater reproducibility.

An alternative matrix which functions essentially as an ion exchanger is hydroxyapatite. Hydroxyapatite matrices employ crystalline calcium phosphate as the ion exchanger and can bind acidic or basic proteins. A more detailed description of hydroxyapatite chromatography can be found in Gorbunoff (1990).

b. Flow properties

Although a fast flow rate can reduce the time for the chromatography experiment, initially it is often better to utilize a matrix with a moderate flow rate in order to be sure that the protein solution has an adequate opportunity to equilibrate and adsorb to the ion exchanger. A matrix with a higher flow rate may subsequently be tried, but remember that faster chromatography flow rates can reduce the resolution of peaks.

c. Capacity

The loading capacity of the matrix is an estimate of how much protein can bind to a given volume of matrix material. Knowing the capacity of an ion exchange matrix is useful in determining how much matrix to use for the quantity of material to be purified. In a situation where a high level of resolution is desired it is better to utilize only 10 - 20% of the available capacity. Note that the capacity of a matrix (defined in milligram protein per milliliter of matrix material) is usually less for a high molecular weight protein than that for a low molecular weight protein. This is because a low molecular weight protein will be able to access more ion exchange sites which are recessed in pores of the matrix.

d. Swelling properties

Some matrices can be compressed when placed under pressure or may swell and shrink under different ionic conditions (for example, Sephadex). This may require that a matrix be removed from a column for regeneration and repacked for the next separation. A propensity to swell can complicate the handling and operation of an ionic exchange column. Newer matrix materials largely overcome these concerns.

e. pH Stability

If the protein application or elution conditions require the use of extremes in buffer pH, make sure to select a matrix which will withstand these conditions. In addition, a weak ion exchanger can be only partially ionized at pH extremes;   therefore, use a strong exchanger if the experimental conditions are at either end of the pH scale.

Table 9.3
Commercially Available Ion Exchangers

Supplier	Name	Type	Matrix	Loading Capacity mg/ml	Flow Rate cm/min*	pH Stability
Pharmacia	DEAE Sepharose Fast Flow	Weak anion	X-linked Agarose	3-110	12.5	1-14
"	DEAE Sepharose CL-6B	"	"	2-170	1.7	2-14
"	DEAE Sephacel	"	Beaded Cellulose	10-160	0.17	2-12
"	DEAE Sephadex A-50	"	X-linked Dextran	2-110		2-9
BioSepra	DEAE Trisacryl M	"	Synth. Polymer	80-90	3	1-11
Bio-Rad	DEAE Bio-Gel A	"	X-linked Agarose	45	>0.3	2-9.5
Pharmacia	CM Sepharose Fast Flow	Weak cation	X-linked Agarose	15-50	12.5	2-14
"	CM Sepharose CL-6B	"	"	10-120	2	2-14
BioSepra	CM Trisacryl M	"	Synth. Polymer	90-100	3	1-11
Bio-Rad	Bio-Rex 70	"	"		0.4-15	5-14
"	CM Bio-Gel A	"	X-linked Agarose	45	>0.3	4.5-10
Pharmacia	CM Sephadex C-50	"	X-linked Dextran	7-140		6-10
"	Q Sepharose Fast Flow	Strong anion	X-linked Agarose	3-120	6.7-11.7	2-12
"	QAE Sephadex A-50	"	X-linked Dextran	1.2-80		2-10
"	SP Sepharose Fast Flow	Strong cation	X-linked Agarose	60	12.5	3-14
"	SP Sephadex C-50	"	X-linked Dextran	8-110		2-10
BioSepra	SP Trisacryl M	"	Synth. Polymer	100	6	1-11

* cm/min = ml/min•cm^2 column cross-sectional area

3. Preparing the Matrix and Packing the Column

Three preparation steps are necessary before a protein solution can be applied to an ion exchange column: (1) the ion exchange matrix must be swollen if it is shipped in a dry state, (2) the matrix must be poured and packed into the column, and (3) the ion exchange matrix must be brought to final equilibration with the column buffer. If the separation is to be performed at 4°C (to minimize enzyme activity losses), all materials should be cooled and stored at 4°C for column packing and subsequent steps.

a. Swelling the Matrix

Matrices which are supplied as dry powders must be hydrated with the column buffer before packing the column. A gram of dry matrix may expand to between 5 and 50 ml of swollen material.

1) Incubate the matrix with about 10 volumes of column buffer (i.e. 10 g matrix + 100 ml buffer).

2) Hydrate by boiling for one to several hours or incubate for several hours to several days at room temperature (consult instructions from supplier). Take care to add more buffer as needed if hydration is by boiling.

3) Swirl the matrix solution several times during the course of swelling to ensure complete hydration.

4) Change the buffer several times during swelling to allow for better matrix equilibration.

Stirring the matrix with a magnetic stir bar is not recommended since the stir bar may shear the matrix into smaller particles ('fines').

b. Packing the Column

Prior to packing the column, the matrix should be equilibrated with the buffer. Significant time can be saved, since sometimes tens of column volumes of washing can be required to achieve adequate equilibration.

In order to pack the column, the matrix slurry should contain one volume of matrix to one volume of buffer.

The column dimensions should be chosen according to the degree of resolution desired. For high resolution, a height:width ratio of 4:1 or 5:1 is often recommended whereas for a less demanding separation, a ratio of 2:1 is generally adequate.

Packing the ion-exchange matrix into a column must be done carefully. Uneven packing can cause poor protein separation or poor reproducibility of the separation, so the matrix must pack into the column in a single step. To provide adequate volume for column packing in a single step, a column reservoir may be used (Fig. 9.3).

1) Wash the matrix in a beaker with the column buffer. Swirl the slurry for a minute, then test the buffer pH. If the pH has changed from that of the original column buffer, further equilibration is needed: decant most of the buffer, add new column buffer, and repeat the slurry swirling and pH test. By partially equilibrating the matrix in this batch fashion, time-consuming washing of the column can be reduced.

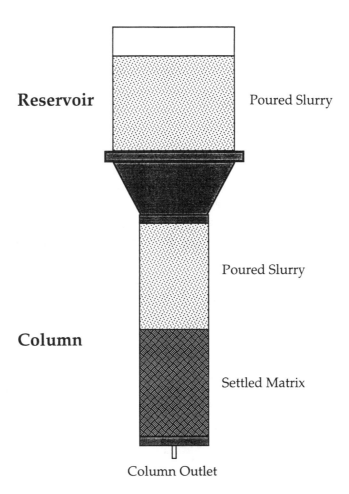

Reservoir   Poured Slurry

Poured Slurry

Column

Settled Matrix

Column Outlet

Figure 9.3. Using a column reservoir.

2) If running the column at room temperature, degas the
   buffer and matrix slurry solutions (apply a vacuum for
   one hour). This step will reduce the likelihood of small
   air bubbles being trapped in the matrix during packing.
   Swirling the matrix during degassing will release air
   bubbles trapped in the matrix.

3) Add a small amount of buffer to the column.

4) Open the column outlet to allow some of the buffer to pass, then close the outlet. This will remove air from the dead space at the bottom of the column.

5) Swirl the matrix solution, then pour into the column down a glass rod, making sure that air bubbles do not become trapped in the matrix.

6) Open the column outlet and add more buffer as the matrix packs.

7) A column adapter can be attached as the matrix is packing. Make sure that no air bubbles enter the system when manipulating the chromatography setup; otherwise, flow rates will be reduced or arrested (consult manufacturer's instructions for fitting the adapter).

The column may also be tilted slightly while the matrix slurry is poured, particularly if the column is a long one. Note that some matrices (such as Sephadex) pack well at low flow rates while many high performance matrices (for example, Sepharose Fast Flow) require high flow rates. Consult the manufacturer's instructions.

c. Equilibrating the Matrix

After packing, the matrix material should be washed further with buffer in order to complete the packing process and to bring the matrix into final equilibrium with the buffer pH and ionic conditions for the ensuing experiment. Washing with several column volumes is generally sufficient to achieve equilibration with most matrices. To be sure that the matrix has been equilibrated, the pH and conductivity of the applied buffer is compared to that of the eluent.

4. Applying the Sample

a. Preparing the Sample for Application

A turbid protein solution should be clarified by ultracentrifugation or by filtration through a 0.45 μm pore size filter. Be sure to check that the protein of interest is not lost during this clarification step.

The sample should be in a low ionic strength buffer (preferably below 50 mM). To change the buffer, use gel filtration chromatography (Chapter 10), dialysis, or ultrafiltration (Chapter 4). Alternatively, dilute the protein solution to reduce the ionic strength.

Note: If the sample protein concentration is too high, the chromatography may suffer due to extreme local pH or salt concentration fluctuations. Protein concentrations above 10 mg/ml should generally be avoided.

b. Sample Application

1) Drain the column until the buffer reaches the surface of the matrix bed (Fig. 4) and close the column outlet.

2) With a pipet, apply the sample gently to the bed surface.

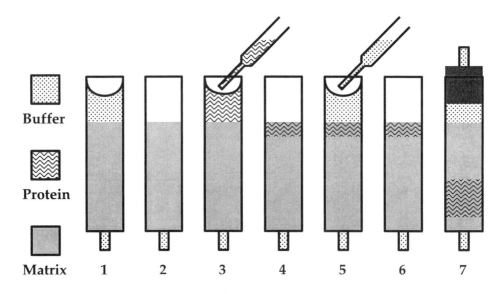

Figure 9.4. Applying the sample to the matrix bed.

3) Open the column outlet until the sample has entered the matrix, then reclose the column outlet.

4) Gently apply some buffer to the bed surface.

5) Open the column outlet so the buffer enters the matrix (carrying with it any remaining sample solution), then close the outlet when the liquid reaches the bed surface.

6) Again add buffer gently to the bed surface, then hook up buffer reservoir.

7) The column is now ready for washing and protein elution.

Note: Care must be taken not to create wells in the matrix when applying sample or buffer solutions since this may cause uneven protein separation. A good strategy for liquid application is to run the pipet slowly around the perimeter of the column wall during application so an even stream of liquid flows onto the bed surface. Use of a flow adapter can simplify the process of sample application (see manufacturer's instructions).

5. Washing and Eluting the Protein of Interest

Washing the column removes unbound protein before eluting any bound protein. In general, 3 to 10 column bed volumes of buffer are adequate for column washing. It is useful to monitor the column eluent for protein concentration or optical density during the wash and elution steps. When minimal quantities of protein are found in the wash flow-through, protein elution may begin.

Two types of elution are possible with an ion exchange column: step elution or gradient elution. In a step elution, discrete increases in ionic strength (for example 0.1 M NaCl, then 0.5 M NaCl, and finally 1.0 M NaCl) cause protein elution (Fig. 9.5). A gradient elution removes proteins by a gradual increase in ionic strength (such as from 0.02 M to 1.0 M NaCl, see Fig. 9.6). Whereas a gradient elution generally gives a more complete separation of protein peaks, a step gradient is frequently a simpler and more rapid procedure. However, a step gradient may result in the simultaneous elution of several protein peaks due to a large ionic strength increase.

We recommend using a gradient elution when a protein separation is being developed or when separation of closely-spaced protein peaks is a concern. A step gradient should be used for a preliminary experiment or after the elution characteristics of the protein of interest have been established.

9

a. Step Elution (Fig. 9.5)

1) Wash column with 3 to 10 bed volumes of first elution
   buffer (for example, 20 mM Tris-HCl, 0.1 M NaCl).

2) Allow buffer to reach the bed surface before applying a
   new elution buffer. This will ensure that the salt
   concentration increases in a well-defined manner.

3) Apply 3 to 10 bed volumes of the second elution buffer
   (e.g. 20 mM Tris-HCl, 0.5 M NaCl), followed by any
   further elution buffers.

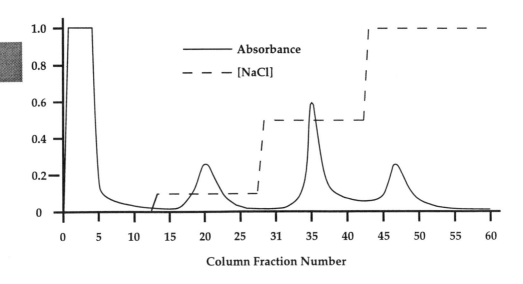

Figure 9.5. Column profile after step elution.

b. Gradient Elution (Fig. 9.6)

Total volume for a gradient elution should be approximately 5 to 10 bed volumes.

1) The lower ionic strength buffer (20 mM Tris-HCl, 0.02 M NaCl) is put into the gradient maker on the side closer to the column. The higher ionic strength buffer (20 mM Tris-HCl, 1.0 M NaCl) is placed into the other side of the gradient maker. The container for each buffer should be identical and the height of the buffer solutions should be equivalent. Make sure that no air bubbles block the connection between the two buffer chambers. Activate the magnetic stirrer in the lower ionic strength side of the gradient maker and attach the gradient maker to the column (Fig. 9.2).

2) Activate fraction collector and recorder.

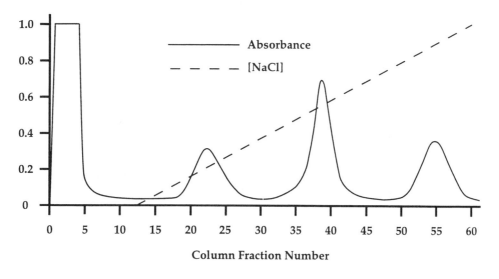

Figure 9.6. Column profile after gradient elution.

6. Regenerating and Storing the Matrix

Following a protein separation, the ion exchange column must be cleaned, and antimicrobial agents should be added prior to storage. Column regeneration removes tightly bound contaminants and prepares the matrix for future separations. Regeneration is possible in the column if the matrix is stable to swelling caused by ionic strength fluctuations (Sephadex ion exchangers should be removed from the column before regenerating). If a matrix cannot resist swelling, the matrix must be removed from the column and repacked after regeneration. The matrix should also be removed from the column for regeneration if significant cell debris accumulates during the purification.

Most ion exchange matrices are regenerated with high concentrations of salt (for example, 1 - 2 M NaCl). More tenacious contaminants such as lipids are often removed with low concentrations of a basic solution (for example, 0.1 M NaOH) or with a detergent (see Table 9.4). Consult the manufacturer's instructions before regenerating an ion exchange matrix.

Table 9.4
Regeneration of Ion Exchange Matrices

Ion Exchange Matrix	Standard Regeneration	Harsher Regeneration Conditions
Sephadex	2 M salt	0.1 M NaOH or nonionic detergent
DEAE Sephacel	2 M salt	0.1 M NaOH or cationic detergent
Sepharose	1 M salt	0.5-2.0 M NaOH
Trisacryl	1-2 M salt	0.1 M NaOH or nonionic detergent
Bio-Rex	0.5 M NaOH	

- Anion exchangers can be stored in a buffered solution containing 0.002% hibitane (chlorhexidine) and cation exchanges can be stored in buffers containing 0.02% sodium azide or 0.005% merthiolate (Thimerosal or ethyl mercuric thiosalicylate). **Warning:** do <u>not</u> mix sodium azide with merthiolate or other heavy metal cations under danger of explosion (Rozycki & Bartha, 1981).

- Some matrices can be autoclaved to prevent microbial growth.

7. Discussion

When problems occur in ion exchange chromatography, they can usually be attributed to one of a limited number of causes. The following list outlines possible causes for problems which may occur and makes suggestions for remedies.

a. The column flow rate is slow
  Solutions
  1) Column outlet is partially closed. Reopen the column outlet.
  2) Air bubbles in tubing are restricting buffer flow. Clear bubbles by increasing column pressure and flicking tubing with a fingernail. Be sure to degas buffers and avoid introducing air bubbles into system.
  3) A precipitate is blocking the top of the matrix. To remove the precipitate, scrape off the top layer of the matrix. "Repack" the top of the matrix by stirring up the first one or two centimeters of matrix material and allowing it to settle before continuing with the elution. Filter the sample more carefully in the future or use detergents during chromatography.
  4) The column support is clogged. The support must be removed and cleaned, possibly at the expense of the sample in the column.
  5) The ion exchange matrix is compressed. This may occur if a matrix (such as Sephadex) can swell due to changes in ionic strength. See Section B.2, "Selecting an Ion Exchange Matrix."
  6) The column is too narrow. For a fine mesh matrix, a height:width ratio of 2:1 may be adequate for good resolution.
  7) Fines are clogging the column. Consult manufacturer's instructions about removing fines.

8) The pump tubing is leaking. Inspect the tubing and replace if necessary.

9) The outlet tubing is clogged. Remove the tubing and clean or replace.

10) Microbial growth has occurred in the matrix. See Section B.6, "Regenerating and Storing the Column."

b. The protein does not adsorb to the matrix.

Solutions

1) The starting ionic strength is too high. Change the starting buffer to one of lower ionic strength. In addition, a trial elution experiment with solutions of different ionic strength can be designed similar to the pH experiment described in Section A.2.

2) The column pH may be interfering with adsorption (see Section A.2). Remember that a calculated isoelectric point may be considerably different from an experimental one.

3) The column may not be properly equilibrated. See Section B.3, "Preparing the Matrix and Packing the Column."

4) The column has not been properly cleaned. See Section B.6, "Regenerating and Storing the Matrix."

c. The yield of purified protein is low.

• Most successful ion exchange purifications have yields of 60-80%. The following explanations can account for lower yields.

Solutions

1) The protein remains adsorbed to the column matrix material. Higher ionic strength concentrations, a more active counter-ion (see Table 9.1) or detergent (Chapter 1) elutions can be tried.

2) The buffer pH is not appropriate. If the buffer pH is significantly different from the pI of the protein of

interest, the protein may bind too tightly to the matrix (see Section A.1).

3) The sample has precipitated. This may be caused by a pH too near the pI (Section A.1), the lack of salt in the buffer or excess dilution of sample.

4) The sample was lost during filtration prior to loading on the column. Be sure that the protein does not stick to the filter material when clarifying a protein solution by filtration.

5) An essential protein or cofactor has been lost during chromatography. To test for the loss of a cofactor or protein, fractions can be remixed and assayed for activity.

6) Proteases may have digested the protein of interest. See Chapter 1, Section VII.

7) Microbial growth may have occurred in the column. See a.10 above.

d. The protein resolution is poor.

Solutions

1) The buffer flow rate may have been too fast, reducing resolution. Try a slower flow rate.

2) The column may have been too short to permit adequate peak separation. Try a longer column.

3) Too much protein may have overloaded the column. Try loading less protein.

4) The gradient slope may have been too steep. Try a shallower gradient.

5) Some protein remixing occurred after leaving column. Be sure to minimize the volume between the column support and the fraction collector (Section B.1).

6) The column is poorly packed. Uneven packing, presence of fines, and improper sample application and elution can lead to poor resolution. See Sections B.3-5.

7) The column is not properly equilibrated. See Section B.3, "Preparing the Matrix and Packing the Column.."

8) Microbial growth may have occurred in the column. See Section B.6, "Regenerating and Storing the Column."

e. The elution profile is not reproducible.
   Solutions
   1) Column equilibration is not complete. See Section B.3.
   2) The pH or ionic strength conditions are not the same as those from the previous experiment. This may be especially troublesome in a gradient elution, so measurement of the eluting ionic strength (or pH) is worthwhile.
   3) The sample precipitated on the matrix. Clean the column to remove any substances which may trigger precipitation and filter the sample.
   4) The sample changed during storage. Retest the protein activity and alter storage conditions if necessary.

f. The sample loses activity when purified.
   Solutions
   1) A cofactor or part of a protein complex has been lost. Remix fractions and test for activity.
   2) The protein is not stable in the elution buffer. Investigate alternative elution conditions.
   3) Microbial growth in the matrix is altering the protein during chromatography. See Section B.6, "Regenerating and Storing the Column."

g. The sample gains activity upon purification.
   Solutions
   1) The assay conditions are different. Be sure no changes have been made in the assay.
   2) An inhibitor has been removed during chromatography. If desired, remix fractions to identify the fractions containing the inhibitory substance.

II. Concentrating a Protein Solution

Ion exchange chromatography can be used to concentrate a protein solution. Since the isoelectric point of most proteins is below pH 8, most proteins will adsorb to an anion exchange matrix at this pH at low ionic strength. A brief wash at high ionic strength will elute nearly all of the bound proteins, giving a simple, rapid concentration step. Batch chromatography may be a useful procedure for protein concentration. The yield of a typical ion exchange concentration step is 60 - 80%, equivalent to what might be expected from an ammonium sulfate precipitation (Chapter 4). Ultrafiltration methods (Chapter 4) can achieve protein concentration yields of 95%. A sample ion exchange concentration protocol follows:

1. Prepare an ion exchange matrix (use DEAE Sepharose CL-6B for example) and pack the column (see Section I.B.3. above) or proceed with batch chromatography (see Section III below). Use one milliliter of matrix material for every 20 mg of protein to be concentrated.

2. Equilibrate the matrix with 20 mM Tris-Cl (pH 8.0), 20 mM NaCl.

3. Apply the sample.

4. Wash the column with 2 bed volumes of 20 mM Tris-Cl, 20 mM NaCl.

5. Elute the sample with 20 mM Tris-Cl, 1 M NaCl.

III. Batch Chromatography

Batch chromatography is a process in which the protein binding, washing, and elution steps are carried out without confining the ion exchange matrix to a column. In general, the matrix is incubated in a beaker or bottle with stirring or rolling, and the buffer is removed by filtration or centrifugation. This procedure is a simple and rapid alternative to column chromatography when the level of resolution of the protein separation is not of paramount concern. Batch chromatography is also an effective method for protein concentration (see Section II above).

1. Equilibrate the matrix in a beaker or Erlenmeyer flask, swirling the matrix frequently to ensure complete equilibration (see Section I.B.3. above).

2. Remove most of the buffer by decanting, leaving enough buffer so that the slurry can still be swirled.

3. Apply the protein sample.

4. Allow the sample to adsorb by incubating for one hour, swirling the slurry several times during the hour.

5. Filter the slurry, taking care not to allow the ion exchange matrix to dehydrate.

6. Wash the matrix gently on the filter with the original buffer solution.

7. Replace the matrix into the beaker or flask and resuspend with a buffer of higher ionic strength.

8. Filter the slurry and test the eluted buffer for presence of the desired protein.

9. If necessary, add an even higher ionic strength buffer to the slurry and continue until the protein of interest elutes from the ion exchange matrix.

IV. Suppliers

Ion Exchange Media	Bio-Rad, BioSepra, Pharmacia, Sigma
Column	Bio-Rad, Pharmacia, Pierce, Sigma, VWR
Tubing	Bio-Rad, Pharmacia
Fraction Collector	Beckman, Bio-Rad, Pharmacia
UV Detector	Beckman, Bio-Rad, Pharmacia, VWR
Chart Recorder	Bio-Rad, Pharmacia, VWR
Peristaltic Pump	Bio-Rad, Pharmacia
Gradient Maker	Bio-Rad, Pharmacia, Sigma

9

V. References

Giri, L. 1990. Meth. Enzymol. 182: 380-392. Chromatofocusing.

Gorbunoff, M.J. 1990. Meth. Enzymol. 182: 329-339. Protein Chromatography on Hydroxyapatite Columns.

Li, C.M. and T.W. Hutchens. 1992. Meth. Mol. Biol. 11: 237-248. Chromatofocusing.

Pharmacia LKB Biotechnology. 1991. Ion Exchange Chromatography: Principles and Methods. Uppsala, Sweden.

Roe, S. 1989. Separation Based on Structure. pp. 175-244 in Protein Purification Methods: A Practical Approach. E.L.V. Harris and S. Angal, eds. IRL Press, New York.

Rozycki, M. and R. Bartha. 1981. Appl. Env. Microbiol. 41: 833-836. Problems Associated with the Use of Azide as an Inhibitor of Microbial Activity in Soil.

Scopes, R.K. 1994. Protein Purification: Principles and Practice, 3rd ed. Springer-Verlag, New York.

9

Chapter 10

Gel Filtration Chromatography

I. Protein Purification
  A. Introduction
  B. Methods

II. Exchanging the Buffer of a Protein

III. Suppliers

IV. References

10

I. Protein Purification

A. Introduction

Gel filtration chromatography is an important method in the repertoire of the protein purifier (Prenata, 1989; Stellwagen, 1990). Other names for this method include size exclusion, gel exclusion, molecular sieve, and gel permeation chromatography. This separation method is unique in fractionating without requiring protein binding, thus significantly reducing the risk of protein loss through irreversible binding or protein inactivation. Gel filtration is also valuable in permitting the transfer of a protein into a different buffer or one of lower ionic strength (see Section II).

When should gel filtration be used in a protein purification procedure? There is no universal answer to this question, yet certain steps in a purification lend themselves well to gel filtration. If the protein is relatively large (for example, over 100 kDal), gel filtration could be used as an initial step in the purification. With the proper filtration matrix, the protein would emerge from the column in the initial protein fractions, and all smaller proteins would still remain in the column, allowing a considerable purification. On the other hand, if an ion exchange column has already been employed to partially purify the protein, gel filtration would be an ideal purification step since the principles of protein separation for the two methods are complementary. In this case, a narrow fractionation range matrix would be chosen in which the protein elutes near the middle of the fractions, permitting a finer separation from proteins of similar sizes. Alternatively, if the molecular weight of the protein is not known, a matrix should be chosen which has a wide fractionation range. Finally, if a protein is already pure, gel filtration can play a role in determining whether the active form is a monomer or oligomer. Thus, gel filtration chromatography can serve a useful function at various stages of protein purification.

1. Gel Filtration Principles

Gel filtration chromatography separates proteins according to their size. The gel filtration matrix contains pores which permit the buffer and smaller proteins to enter but which exclude larger proteins and protein complexes. Therefore, larger proteins migrate around the matrix particles and elute from the column before the smaller proteins (Fig. 10.1). The largest proteins emerge from the column first since they have the smallest volume to pass through before reaching the end of the column. Medium-sized proteins can enter the larger size pores in the matrix, and so they reach the end of the column later. Small proteins are able to enter all the pores, and they have the largest volume to pass through before emerging from the column last.

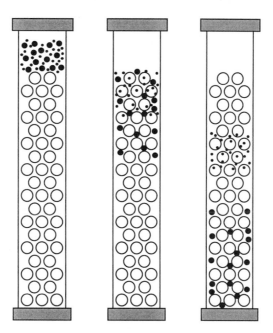

Figure 10.1. Schematic representation of gel filtration. Open circles represent the matrix, closed circles represent proteins.

The objectives of the gel filtration separation are essential in guiding the selection of the purification approach. These objectives can be high protein resolution, minimal chromatography time, minimal sample consumption (for an analytical application), or maximal reproducibility of the results. Carefully defining the objectives will be helpful in selecting the appropriate chromatography equipment. If the chief objective is to achieve maximal resolution between proteins of different sizes, the gel filtration column must be long and narrow (column height is generally 20 to 40 times the column diameter for high resolution separations). These proportions are required for adequate filtration since no binding occurs on the matrix. Some exploratory experiments may be valuable for comparing different types of matrices, which sometimes result in different degrees of separation. In addition, smaller size matrix particles and a lower flow rate contribute to higher resolution separations. However, if speed is important, some compromise in resolution may be required; otherwise, careful attention to selecting an appropriate high resolution matrix with a relatively high flow rate will be necessary. If the overriding concern is to conduct a gel filtration experiment with minimal sample amounts, an automated high resolution system such as the Pharmacia SMART™ System should be considered. Finally, if run-to-run reproducibility is required, investing in a prepacked column and automating the chromatography equipment may be advisable.

Other uses have been developed for gel filtration chromatography:

- **Molecular weight determination**: Gel filtration can provide an estimation of a protein's molecular weight, since proteins normally pass through a column without physically interacting with the matrix. However, any estimation of a molecular weight contains the assumption that the protein is globular. This is because most protein molecular weight standards are globular themselves. Thus, a rodlike protein will be assigned a relatively high molecular weight if excluded from the matrix pores or a low molecular weight if the protein enters the pores and becomes constricted. A common strategy for removing protein geometry concerns from a molecular weight determination is to treat the protein with guanidinium hydrochloride or urea prior to chromatography (Ansari and Mage, 1977). Under these denaturing conditions, the protein tertiary structure is reduced to a random coil and a more representative molecular weight can be determined. However, information on the number of subunits in an oligomeric complex is lost.

- **Removal of low molecular weight impurities.** Rapid separation of undesired radioactivity, proteolytic protein fragments, cofactors, or low molecular weight proteins can be achieved. See Section II.

- **Separation of proteins from dimers and other oligomers**.

10

- **Exchanging the buffer of a protein**. If a matrix is chosen which completely excludes the protein of interest, the protein can be eluted into a new buffer (see Section II). The column is simply equilibrated with the new buffer before the protein is applied. This procedure is also used for "desalting" a protein: a protein in high ionic strength buffer is exchanged into a low ionic strength buffer. This is useful before application of the protein to an ion exchange column (Chapter 8) or if the protein is less active in the high salt buffer.

- **Studying protein-ligand binding.** See Hummel & Dreyer (1962) or Ackers (1973).

2. A Sample Separation Protocol

A sample separation will serve to highlight the steps involved in running a gel filtration experiment:

A protein of molecular weight 50,000 daltons is to be resolved using 2 ml of a 5 mg/ml protein solution. The protein is known to be stable in pH 8 Tris buffer.

a. **Choose the matrix.** For a 50 kDal protein, a matrix such as Sephacryl S-200 should work well (from Table 1). Sephacryl S-200 has a fractionation range of 5 kDal to 250 kDal, so a 50 kDal protein is in the proper size range. In addition, if the 50 kDal protein exists as a dimer or oligomer, the protein should still be reasonably well-resolved with this matrix. See Section A.3 for more details.

b. **Determine the column dimensions.** Two calculations must be made: the appropriate matrix volume to use and the column geometry. If the sample volume is to comprise 2% of the matrix volume (see Section B.3 for more details), 100 ml of matrix will be adequate (2 ml sample volume is 2% of 100 ml). For a well-resolved separation, the column height should be 20 to 40 times the column diameter. Thus, a 100 ml matrix could be poured into a 1.5 ml diameter column. The bed height should be approximately 56 cm (Volume = $\pi r^2 h$; 100 ml = 100 cm^3 = $\pi \times (0.75$ cm$)^2 \times$ h).

Scopes (1994) suggests a formula for column diameter as $(m/10cm)^{1/3}$, with m being the amount of protein in mg, and the column length as 30 x diameter.

c. **Determine the experimental buffer conditions**. In this experiment, a pH 8 Tris buffer would be appropriate. Be sure that the protein is stable in the experimental buffer and that any necessary cofactors, cations, or reducing agents are present. The temperature at which the chromatography experiment will be conducted must also maintain protein stability. In addition, a moderate ionic strength salt (0.2 M NaCl, for example) would help to insure against possible protein interactions with the matrix.

d. **Set up the chromatography equipment**. Consult Section B.1 for information about gel filtration chromatography components.   Obtain and assemble the apparatus in the configuration shown in Fig. 10.2.

e. **Prepare the matrix and pack the column**. See Section B.2 for instructions about readying the column for a chromatography run.

f. **Run the chromatogram**. Section B.3 describes how to introduce the sample and buffer into the column.

10

g. **Test the fractions for the protein**. The protein of interest can be detected by activity, immunological methods, or molecular weight. Enzyme activity tests which are unique to the protein are most useful in protein purification. Immunoblotting (Chapter 8) employs an antibody to detect the protein; immunoblotting is generally prone to more interpretation errors than an activity assay and should be utilized with care.   SDS-Polyacrylamide gel electrophoresis (SDS-PAGE;   Chapter 5) can be used to determine the molecular weight of a protein. Electrophoretic analysis can be used to narrow the number of fractions to be tested by a cumbersome or expensive activity assay.

h. **Analyze the results**. If the protein elutes early in the run, the peak is likely to be sharper;   however, contaminating protein peaks may elute nearby.   Conversely, a protein eluting later during the gel filtration may be better resolved from neighboring peaks but the peak itself is likely to be broader. Consult Section B.5 for suggestions if the chromatography results should be improved.   Chapter 9 (Ion Exchange Chromatography) or Chapter 11 (Affinity Chromatography) offer further purification steps.

10

3. Selecting a Gel Filtration Matrix

- The choice of a gel filtration matrix should be based primarily on the molecular weight of the protein to be separated. Objectives other than protein separation will also influence the matrix selection, and these are discussed in Section 1 above. The molecular weight should be in the middle of the matrix fractionation range for optimal separation (consult Table 10.1 for characteristics of some matrices). Note that employing a matrix with a narrower fractionation range will allow separations with a better resolution. If no information about the protein is available, a matrix with a wide fractionation range should be chosen (for example, Sephacryl S-300). For a more complete listing of matrices, consult Patel (1993) or the annual listing of suppliers published by *Science* or *Biotechnology*.

- Gel filtration matrices are often made of either cross-linked dextran, polyacrylamide, or agarose beads. The dextran and polyacrylamide matrices separate proteins of small to moderate molecular weights, while the agarose matrices have larger pores and are able to separate much larger protein complexes. Some agarose matrices have the disadvantage of being compressible under pressure, so care must be exercised when attaching a peristaltic pump to these matrices.

- Matrix flow rates can also be considered when selecting a gel filtration matrix. Faster flow rates will allow a separation to proceed more rapidly; however, slower flow rates sometimes offer better resolution of peaks. Some matrices are available in different grades (such as coarse, medium, and fine) to allow the investigator latitude in defining the balance between speed and resolution. Coarse grades provide higher flow rates whereas finer grades permit better separation of peaks. Some trial and error experiments may be required to determine which flow rate can be tolerated without compromising the purification step. Newer "High Resolution" matrices can be run at high flow rates without compromising resolution.

Table 10.1
Commercially Available Gel Filtration Matrices

Supplier	Name	Gel Type	Fractionation Range, kDal	Flow Rate (cm/hr)*
Bio-Rad	Bio-Gel P-6DG	Polyacrylamide	1-6	15-20
	Bio-Gel P-30	"	2.5-40	6-13
	Bio-Gel P-60	"	3-60	3-6
	Bio-Gel P-100	"	5-100	3-6
	Bio-Gel A-5m	Agarose	10-5000	7-25
	Bio-Gel A-15m	"	40-15,000	7-25
	Bio-Gel A-50m	"	100-50,000	5-25
Pharmacia	Sephacryl S-100 HR	X-Linked Dextran/Bis-acrylamide	1-100	20-39
	Sephacryl S-200 HR	"	5-250	20-39
	Sephacryl S-300 HR	"	10-1500	24-48
	Sephadex G-25	Dextran	1-5	2-5
	Sephadex G-50	"	1.5-30	2-5
	Sepharose CL-6B	Cross-Linked Agarose	10-4000	30
	Superdex 75	Agarose/ Dextran	3-70	7-50
	Superdex 200	"	10-600	7-50
	Superose 6	Cross-Linked Agarose	5-5000	30

*cm/h = ml/hr•cm^2 column cross-sectional area; note that the cited rates are estimates.

Other companies offering gel filtration matrices are listed in Section III.

B. Methods

1. Equipment

A basic gel filtration chromatography setup (Figure 10.2) requires only a column with tubing attached to a fraction collector. A reservoir automates the task of adding buffer to the column, and a peristaltic pump can regulate and speed the buffer flow rate with increased pressure. Finally, a UV detector and a chart recorder relieve the need to test all the collected fractions for protein content.

More sophisticated integrated chromatography equipment is also commercially available, including high pressure systems which permit more rapid separations. The researcher should be guided by the protein separation requirements in determining which system to purchase. Due to the considerable time requirements of a standard gel filtration chromatography run (12 hours is not uncommon), the significant investment for a high pressure chromatography system may be necessary when the available time for protein purification is limited. For most applications, however, standard chromatography equipment produces excellent results. The discussion in this text will center around the standard 'low-pressure' chromatography equipment and methodology which are most commonly used in protein purification laboratories.

10

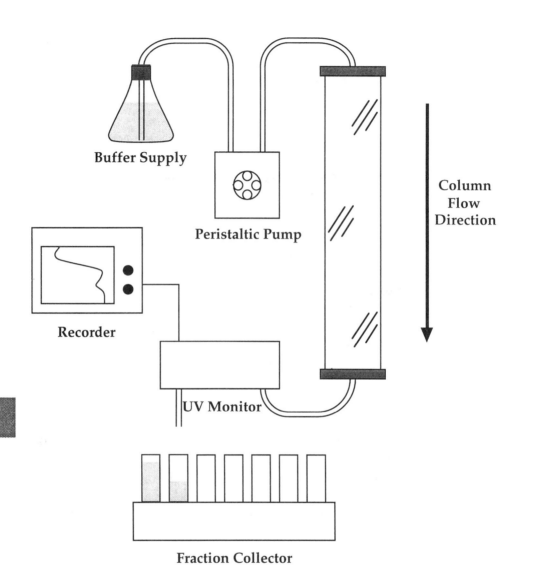

Figure 10.2. Gel filtration chromatography equipment.

a. **Column**. The gel filtration column is similar to the column used for ion exchange chromatography (Chapter 9). The column should have a porous support, and the 'dead space' after the support should be minimal to reduce any sample remixing after the chromatography (< 0.1% of the column volume is standard). Gel filtration columns are usually long and narrow (the height should be 20 - 40 times the diameter) to permit maximal separation, although the diameter should be at least 1 cm to reduce anomalous effects due to protein interaction with the column wall. If the sample volume requires that an extremely long column be used to achieve adequate separation, more than one column can be attached in series. To attach columns in series, connect the bottom of one column to the inlet of the next column with tubing. In addition, prepacked gel filtration columns are commercially available.

b. **Tubing**. The tubing for connecting the gel filtration components should be flexible. Teflon®, Tygon®, and silicone tubing are often used.

c. **Fraction Collector**. The fraction collector presents a new collection tube at the column outlet after a specified volume has eluted or a given period of time has elapsed.

d. **Buffer Reservoir**. The reservoir stores the elution buffer and is connected to the column by tubing. The buffer reservoir can be a beaker or Erlenmeyer flask.

e. **Peristaltic Pump**. A pump regulates the rate of buffer flow to the column. For matrices which can withstand some pressure, a peristaltic pump is useful for accelerating the column flow rate.

f. **UV Monitor**. An ultraviolet wavelength detector attached to the column outlet permits protein peaks to be directly detected (for example, at 280 nm). The column peak profile is traced on a chart recorder, saving the time and effort of testing each column fraction for protein content.

g. **Chart Recorder**. A chart recorder prints the absorbance profile detected by the UV monitor.

10

2. Preparing the Matrix and Packing the Column

a. Swelling the Matrix

If the matrix has not been hydrated by the manufacturer, the dry powder will require swelling.

1) Add approximately ten parts buffer to one part matrix material.

2) Agitate the slurry on a rotary shaker or swirl the mixture by hand occasionally. Agitation permits uniform hydration of the gel filtration matrix.

3) Swelling can take an hour if the material is boiled or overnight at room temperature. Consult the matrix manufacturer's instructions.

4) When the matrix is not swelled by boiling, degassing is necessary to reduce the risk of air bubble formation in the column. Apply a vacuum to the matrix slurry for an hour, agitating the slurry occasionally by hand or continuously with a rotary shaker.

- Do not agitate the slurry with a magnetic stirrer. Mechanical agitation can break the matrix material into 'fine' particles which can interfere with chromatography. If 'fines' are a concern, they can be decanted after swirling the matrix slurry and allowing the matrix particles to settle. This procedure should be repeated a few times until the liquid phase is visibly less cloudy.

b. Packing the column

- Prior to packing the column, the matrix should be in a thick slurry with the matrix material taking up 3/4 of the volume and the buffer taking up the remaining 1/4. The matrix and column should be at the column operating temperature (usually 4°C).

- If a column is to be used frequently, the use of flow adapters is recommended. Adapters, which are commercially available, can help protect the column bed surface and shield the matrix from large particles in the sample, helping to extend the column life. Instructions for using adapters are available from the suppliers.

1) Add buffer to the column, then close the outlet after some of the buffer has run out. This should remove any air bubbles from the space below the column support.

2) In a single step, open the column outlet and pour the matrix slurry into the column, either along the side of the column or down a glass rod.

10

- Special care must be taken not to introduce air bubbles while the column is packing. If bubbles develop early during packing, a glass rod can be used to stir the matrix and release the air bubble.

- A column extension or funnel may be needed to allow packing of the column in a single step (see Chapter 9, Fig. 9.3). If a column is not packed in a single step, the bed may be uneven.

3) After some of the matrix has settled in the bottom of the column, open the column outlet to hasten the packing, and add more buffer.

4) Wash the column with several column bed volumes to allow stabilization and equilibration. The column can be connected to a pump and reservoir at this time.

- Sephacryl HR and Superose matrices are packed in a two step process using different flow rates. Consult Pharmacia (1991) for packing flow rates.

- DO NOT allow the column matrix to run dry at any time or irregularities in the matrix will develop which can interfere with proper separation.

10

c. Checking the column

- Since air bubbles or uneven column packing can cause a significant loss of resolution during chromatography, examine the column carefully for signs of poor packing.

1) Visually inspect the column for air bubbles or cracks in the packed matrix. A hand-held lamp can be useful for this inspection.

2) For a more careful inspection of column packing, a 2 mg/ml solution of blue dextran can be applied to the column. The applied volume should be 1% of the column bed volume, and the dye should elute in no more than twice the applied volume.

- Blue dextran is easily visible and should pass through the column as an evenly flowing band with no streaking if the column has been packed well. Hemoglobin or ferritin are colored proteins which are recommended in place of blue dextran for Bio-Gel P columns. These markers are useful in defining the extent of protein dilution during a column run.

- Blue dextran can also be used to determine the 'void volume' of the column. The void volume refers to the volume required for elution of a particle which does not enter any of the pores in the gel filtration matrix. Thus, the void volume will define the earliest point at which a protein could possibly emerge from the column, and the volume required for eluting other proteins can be measured relative to this standard.

3. Applying and Eluting the Sample

a. Sample Application (see Fig. 9.4)

- The protein sample should be highly concentrated (10 - 20 mg/ml; consult Chapter 4) and the sample volume should be small (typically 1 - 5% of the column bed volume). An application volume greater than 5% of the bed volume may reduce resolution, while an applied volume of less than 1% is unlikely to improve the separation. In addition, the sample should not be significantly more viscous than the buffer or peak broadening may occur.

- The protein sample should be clarified by filtration (0.2 μm pore size filter) or centrifugation (5 minutes at 10,000 x g) to remove debris which may interfere with the chromatography.

- The chosen buffer should preserve protein activity and prevent nonspecific protein-protein or protein-matrix interactions. In general, a low ionic strength salt (20 to 100 mM) is sufficient to block nonspecific ionic interactions. In some cases, hydrophobic interactions with the matrix are possible in which case a high ionic strength salt should not be used.

1) Elute the buffer until the meniscus reaches the top of the matrix bed, then close the column outlet. NEVER allow the column matrix to run dry or the column must be repoured.

2) Gently apply the sample solution to the top of the matrix.

10

3) Open the column outlet and allow the sample to enter the matrix, then close the outlet.

4) Add a small amount of buffer to the top of the matrix.

5) Open the column outlet until the buffer enters the matrix, then shut the outlet.

6) Apply more buffer to the matrix.

7) Elution can begin. The reservoir and pump can be attached now.

b. Sample Elution

- Buffer is allowed to run through the column until the protein of interest has eluted.

- A peristaltic pump is commonly used to control the column flow rate. When using a pump, do not exceed the maximum operating pressure of the matrix (consult manufacturer's instructions).

- Since gel filtration columns can take many hours to run, the inlet tubing should descend below the level of the column outlet before feeding into the column itself (Fig. 10.3) when not using a peristaltic pump. This arrangement will stop the buffer flow before the column can run dry.

- Yields are typically over 85%.

- Peak resolution generally improves with a slower flow rate.

10

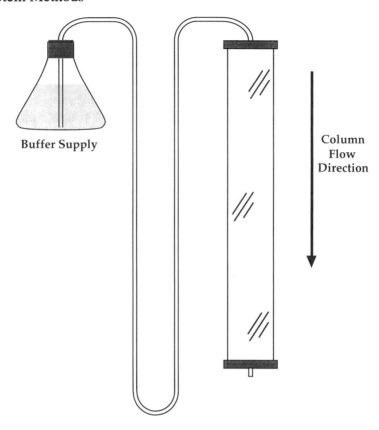

Buffer Supply

Column
Flow
Direction

Figure 10.3. Tubing arrangement to prevent column from
running dry.

- To estimate the molecular weight of a protein (keeping
  in mind that size estimations are relative to globular
  proteins except under denaturing conditions), a standard
  curve can be generated by running protein standards
  over the gel filtration column. Bio-Rad, Pharmacia, and
  Sigma provide protein size standard kits. Note that the
  denaturing conditions often employed in determining
  molecular weight are not compatible with all matrix
  types. The elution volumes of standard proteins plotted
  as a function of the logarithm of their molecular
  weights should generate a straight line. Then, the
  elution volume of an unknown protein can be plotted
  onto this standard curve to give a molecular weight
  (Andrews, 1965).

4. Regenerating and Storing the Column

Since gel filtration matrices are not designed to bind proteins, removing proteins from the matrix should not be a difficult task. In general, a wash with a dilute sodium hydroxide or nonionic detergent solution will remove most bound materials (see Table 10.2 below for some examples). The best insurance against having materials bind is to filter the sample well before applying and to use fresh, filtered buffers for chromatography.

Table 10.2
Regeneration of Some Gel Filtration Matrices

Gel Filtration Matrix	Regeneration Treatment
Bio-Gel P	3% H_2O_2
Bio-Gel A	0.01% diethylcarbonate
Sephacryl HR	0.2-0.5M NaOH or nonionic detergent
G-type Sephadex	Nonionic detergent or 0.2M NaOH
Sepharose CL	0.5M NaOH or 1% nonionic detergent
Superose	0.1-0.2M NaOH

The following procedures can be tried if standard regeneration attempts do not remove a contaminant (Pharmacia, 1991). More efficient washing may be obtained by treating the matrix outside of the column and then repouring the column. Hydrophobic proteins may be removed by eluting the column overnight with an organic solvent (24% ethanol or 30% acetonitrile). Hydrophilic proteins can be eluted with 30-50% acetic acid or by incubating the column overnight with 1 mg/ml pepsin (in 0.5 M NaCl and 0.1 M acetic acid; wash the column well after protease treatment to eliminate traces of pepsin). Nucleic acids may be washed out with TE buffer (10 mM Tris-HCl, 1 mM EDTA, pH 8.0) or can be treated with nuclease. Lipids can be washed off the column by an overnight elution with a nonionic detergent such as 0.2 - 1% NP-40 or Lubrol.

The column should be stored in a buffer containing an antimicrobial agent. Storage in the cold is an added deterrent to microbial growth. Sodium azide (0.02%) is often included in the buffer to prevent the growth of microorganisms. Ethanol (20%) can be used with Sephacryl HR or Sepharose CL. Other antimicrobial agents may be recommended by the matrix manufacturer. Some matrices can also be autoclaved to insure against microbial growth.

10

5. Discussion

The following list outlines problems which can occur in gel filtration chromatography and provides suggestions for overcoming each difficulty.

a. Low column flow rate
1) Column outlet is closed. Reopen the column outlet.
2) Air bubbles in the tubing are restricting buffer flow. Clear bubbles by increasing column pressure and flicking the tubing with a fingernail. Make sure to degas buffers and matrices and to pour columns more carefully.
3) Tubing may be obstructed. Stop buffer flow and exchange tubing. Sometimes an obstruction can be cleared by adding detergent to the buffer. Otherwise, the column must be cleaned and repacked.
4) Precipitate is blocking top of matrix. Scrape the surface of the matrix and remove any residue, then stir up the top one to two centimeters of matrix material and allow to settle before continuing with the elution. Filter the sample more carefully or use detergents during chromatography.
5) Matrix is compressed or is incompletely swollen or contains too many 'fines.' Sephadex matrices can compress if placed under too much pressure. The column must be repacked.
6) Pump tubing is leaking. Inspect tubing and exchange if necessary.
7) Microbial growth in the matrix. See Section B.4, "Regenerating and Storing the Column." A new column must be prepared.

10

b. Unusual peak profile
   1) Poor sample application. Sample application can be practiced with blue dextran (see Section B.2.c, "Checking the column").
   2) Protein adsorbed to the matrix (especially if peaks have a long tail). Include a higher ionic strength salt or some detergent in the buffer. In some cases, a hydrophobic interaction may be involved, in which case the salt ionic strength should be reduced.

c. Poor resolution of desired protein peak
   1) Column is too short. Resolution increases as the square root of column length.
   2) Flow rate is too fast. High flow rates reduce resolution.
   3) Large 'dead space' at bottom of column. In the dead space between end of column and fraction collector, eluted proteins can remix, reducing resolution.
   4) Sample volume is too large. For best resolution, sample should comprise 1 - 5% of column bed volume.
   5) Column is poorly packed. Poor packing can cause strange buffer flow patterns which reduce resolution.
   6) Wrong matrix used. Use a matrix with an appropriate fractionation range. A matrix with a narrower fractionation range will give better resolution. Remember that a nonglobular or denatured protein has different elution characteristics from a globular protein of the same molecular weight.
   7) Wrong matrix grade. Finer matrix grades generally offer better resolution in exchange for longer elution times.

10

d. Poor sample recovery
   1) The sample precipitated. Insufficient or excessive salt concentrations in the buffer can lead to protein precipitation. Verify that the protein is soluble under experimental conditions.
   2) The sample was lost before being loaded. If the protein solution is clarified by filtration or centrifugation prior to column loading, check sample recovery before loading on column.
   3) Protein adsorbed to the matrix. See b.2. above.
   4) Elution conditions are too harsh. Under harsh conditions, protein subunits can dissociate or essential cofactors can be released. Test for protein activity under the elution conditions. If loss of subunits or cofactors is suspected, fractions can be remixed to regain activity.
   5) Proteolysis. Include protease inhibitors in buffer (see Chapter 1).
   6) Microbial growth in the matrix. See a.7 above.

e. Elution profile is not reproducible
   1) The buffer or column composition is not the same as that of the previous experiment.
   2) The sample precipitated. See d.1 and 2 above.
   3) The sample changed during storage. Retest protein activity and alter storage conditions if necessary.

f. Protein activity is lost
   1) A cofactor or part of a protein complex has been lost. Remix fractions and test for activity.
   2) The protein is not stable in the experimental buffer. Investigate alternative buffers.
   3) Microbial growth in the matrix is altering the protein. See a.7 above.

10

II. Exchanging the Buffer of a Protein

Gel filtration chromatography is a mild and rapid method for transferring a protein mixture from one solution to another. Frequently, gel filtration is used to remove a protein to a low ionic strength buffer (for example, an ammonium sulfate precipitation or a peak from an ion exchange chromatography column) in a variation termed 'desalting.' In addition, low molecular weight contaminants such as unbound radioactive isotopes or nucleotides can be removed rapidly from a protein solution.

This method takes advantage of the properties of some matrices to exclude essentially all protein from the pores. Thus, the applied protein emerges from the column in the presence of a pre-equilibrated buffer before the applied buffer or low molecular weight contaminants can pass through the entire column volume. Since the protein does not interact with the column matrix, there is practically no risk of protein denaturation which might occur with ultrafiltration (Chapter 4). Finally, relatively high flow rates and the possibility of gel filtration buffer exchange by centrifugation make this method a rapid step during a protein purification as compared to dialysis (Chapter 4). However, desalting by gel filtrations is not always complete; thus, dialysis is still preferred for more complete salt removal.

- The protein concentration should be less than 30 mg/ml and the sample volume may be as large as 20 - 30% of the column bed volume. Smaller relative sample volumes will be required when removing a high concentration of a buffer component. A short, wide column can be used to achieve a faster flow rate.

- Prepare the matrix and pack a column with a low molecular weight fractionation range matrix such as Sephadex G-25 (Fine Grade) or BioGel P-6DG (see Section I.B.2 above).

- Equilibrate the column with the buffer that the protein will be exchanged into.

- Apply and elute the sample (see Section I.B.3 above).

- The buffer used for elution is not important since the protein will pass through the column faster than any applied buffer.

Prepacked gel filtration columns are commercially available for centrifugation in a low speed benchtop centrifuge (for example, NICK Spin or PD-10, Pharmacia; Bio-Spin, Bio-Rad). These columns can be equilibrated and can exchange buffers for protein solutions up to 0.1 ml in a few minutes.

10

III. Suppliers

Suppliers of Gel Filtration Media and Equipment

Gel Filtration Matrix: Amicon, Bio-Rad, BioSepra, Pharmacia, Sigma, Spectrum, see also Patel (1993)

Column: Amicon, Bio-Rad, BioSepra, Pharmacia, Pierce, Sigma, Spectrum, VWR

Tubing: Baxter, Bio-Rad, Pharmacia, Spectrum, Thomas, VWR

Fraction Collector: Beckman, Bio-Rad, Pharmacia, Spectrum

Peristaltic Pump: Baxter, Bio-Rad, Pharmacia, Spectrum, Thomas, VWR

UV Monitor: Beckman, Bio-Rad, Pharmacia, Spectrum, VWR

Chart Recorder: Baxter, Bio-Rad, Pharmacia, Spectrum, Thomas, VWR

Consult also the annual buying guides from *Science* or *Biotechnology*.

10

IV. References

Ackers, G.K. 1973. Meth. Enzymol. 27: 441-455. Studies of Protein Ligand Binding by Gel Permeation Techniques.

Andrews, P. 1965. Biochem. J. 96: 595-606. The Gel-filtration Behavior of Proteins Related to their Molecular Weights over a Wide Range.

Ansari, A.A. and R.G. Mage. 1977. J. Chrom. 140: 98-102. Molecular-weight Estimation of Proteins using Sepharose CL-6B in Guanidine Hydrochloride.

Hummel, J.P. and W.J. Dreyer. 1962. Biochim. Biophys. Acta 63: 530-532. Measurement of Protein-binding Phenomena by Gel Filtration.

Patel, D. 1993. Chromatographic separation media. pp 49-68 in Biochemistry LabFax, J.A.A. Chambers and D. Rickwood, eds. Academic Press, San Diego.

Pharmacia LKB Biotechnology. 1991. Gel Filtration: Principles and Methods. Uppsala, Sweden.

Prenata, A.Z. 1989. Separation on the Basis of Size: Gel Permeation Chromatography. pp. 293-305 in Protein Purification Methods: A Practical Approach. E.L.V. Harris and S. Angal, eds. IRL Press, New York.

Scopes, R.K. 1994. Protein Purification: Principles and Practice, 3rd ed. Springer-Verlag, New York.

Stellwagen, E. 1990. Meth. Enzymol. 182: 317-328. Gel Filtration.

10

Chapter 11

Affinity Chromatography

I. Introduction

II. Preparing an Affinity Column
   A. Ligand Immobilization on Cyanogen Bromide-Activated Sepharose
   B. Nonspecific Elution Strategies

III. Specialized Techniques
   A. Antibodies
   B. Nucleic Acid
   C. Lectin
   D. Dye Ligand
   E. Immobilized Metal Affinity Chromatography
   F. Hydrophobic Interaction Chromatography

IV. References

11

I. Introduction

Affinity chromatography encompasses a diverse array of purification methods, all based on a specific affinity between a protein and its matrix-bound ligand. The ligand can be biologically specific, such as a peptide, antibody or nucleic acid, or it can exploit nonspecific interactions with lectins, dyes, or hydrophobic moieties. Although separations based on specific protein-ligand affinities tend to be unique for different proteins, our aim is to provide the investigator with protocols describing the most common methods along with some useful guidelines for alternative binding and elution approaches.

Section II provides a general description of the tools required for affinity chromatography followed by a discussion of nonspecific elution strategies. In Section III, more specialized affinity chromatography approaches are discussed. References provided in these sections should assist the reader in entering the purification literature.

In all affinity purifications, the most effective strategy is to bind the protein of interest to a stationary phase and then to desorb it with an eluent specific for that protein (a substrate, peptide, or cofactor, for example). The specificity and mildness of these interactions allows a significant purification in a short time. Nonspecific perturbation of the ligand-protein interaction is required in other cases, though this often results in reduced levels of purity and sometimes in some loss of protein activity. Since the exact physical nature of the interaction between an affinity ligand and the bound protein is often not known, a trial and error approach is required for establishing the most effective purification scheme. This chapter provides suggestions for developing an affinity chromatography strategy during protein purification.

II. Preparing an Affinity Column

An affinity column must be prepared by linking the affinity ligand of interest to an insoluble matrix. For a great many applications, commercially available affinity matrices are convenient and relatively cost-efficient. However, for those applications which require specialized affinity matrices, this section provides details for generating an affinity column.

Coupling an affinity ligand to a matrix can be accomplished using a variety of reagents. Generally, a two-step process is employed involving the attachment of a reactive group to the matrix followed by generation of a covalent bond between the reactive group and the ligand (Fig. 11.1). The investigator must decide on a coupling strategy, which reactive group to use, and which matrix to use.

In this section, we highlight ligand immobilization with cyanogen bromide-activated Sepharose. Most affinity ligands (such as peptides, proteins, lectins, and nucleic acids) can be coupled readily using this method, as discussed below. Other coupling reagents are listed in Table 11.1. The cyanogen bromide procedure is widely used, and an extensive literature attests to successful separations with affinity ligands immobilized by this technique.

11

A.

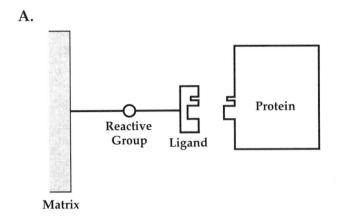

Matrix

B.

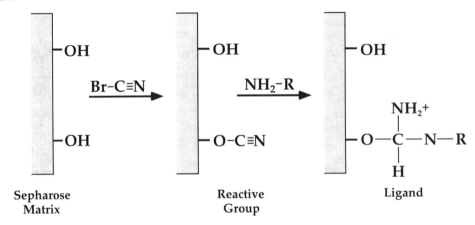

Figure 11.1. Schematic of a matrix with a reactive group, a ligand, and the protein of interest.

A brief discussion of the principles and limitations of ligand immobilization follows:

Choice of coupling strategy. Two general strategies may be employed for coupling a ligand to a matrix:

i) The ligand can be mixed with a matrix containing an activated reactive group, resulting in covalent bond formation between the reactive group (for example, cyanogen bromide) and the specific coupling group on the ligand (i.e. $-NH_2$). By preactivating the reactive group on the matrix, the ligand is spared from having to endure potentially damaging harsh conditions.

ii) The ligand can be mixed with a bifunctional reactive group, and the activated ligand subsequently mixed with a matrix. These procedures are less generally useful (primarily allowing coupling of the ligand to a wider range of matrices) and will not be discussed further.

Choice of coupling reagent. A variety of reagents are available for immobilizing different sites on ligands to various matrices (see Table 11.1 below). Commercially available activated matrices are not expensive and generally do not require handling of toxic chemicals. When no guidelines are available for coupling a given ligand, a good strategy is to adopt a simple, convenient activated matrix to start. If unacceptable binding efficiency or ligand inactivation result, switch to a different coupling reagent which attaches to a different site on the ligand. These and other coupling chemistries are described by Harlow & Lane (1988), Pepper (1992b), and Pharmacia (1993).

Choice of matrix. If a very high molecular weight antigen is to be purified, a large pore size matrix should be chosen. Matrices with large pore sizes must not be too compressible; otherwise, unacceptably slow flow rates may result. For most applications, agarose, cross-linked agarose, polyacrylic, or polyacrylamide beads work well.

11

Table 11.1
Affinity Chromatography Coupling Reagents

Ligand Binding Group	Coupling Reagent	Compatible Matrix	Supplier
amino	CNBr	agarose, cross-linked agarose, polyacrylic, polyacrylamide-agarose	IBF, Pharmacia
amino	glutaraldehyde	polyacrylamide-agarose	IBF, Sepracor
amino	hydroxysuccinimide	cross-linked agarose	Bio-Rad, Pharmacia, Sigma
amino or sulfhydryl	tosyl or tresyl chloride	agarose, cross-linked agarose, polyacrylic	Pierce, Schleicher & Schuell, Sigma
carboxy	1,6-diaminohexane	cross-linked agarose	Pharmacia
amino or carboxy	carbodiimide	agarose, cross-linked agarose, polyacrylic, polyacrylamide-agarose	Bio-Rad, Pharmacia, Pierce
carbohydrate	hydrazide	cross-linked agarose, polystyrene, polyacrylamide	Bio-Rad, Pierce, Sigma

11

A. Ligand Immobilization on Cyanogen Bromide-Activated Sepharose

Cyanogen bromide activated matrices react with primary amino groups to form isourea linkages (Fig. 11.1). Since a small amount of ligand leakage from the matrix can be expected if the column is to be used repeatedly, other activated matrices (for example, hydroxysuccinimide) should be considered for multiple use affinity columns. Formation of the isourea linkage is accompanied with the addition of a charged group, thus necessitating the subsequent use of at least 0.1 M salt in purification buffers to overcome an ion exchange effect. This procedure employs inexpensive reagents and is widely used with proteins, nucleic acids, and lectins.

Protocol
1. Suspend 2 g CNBr-activated Sepharose in 1 mM HCl gently on a sintered-glass funnel. Stir slurry with glass rod for about 20 min until matrix is swollen, then apply vacuum to remove liquid.
   - 1 g dry gel gives about 3.5 ml swollen gel, which can bind 30 - 200 nmol ligand
   - HCl preserves the reactivity of the active group

2. Wash matrix with 3 x 200 ml 1 mM HCl, washing to dryness each time.

3. Wash matrix with 100 ml Coupling Buffer (0.1 M $NaHCO_3$, 0.5 M NaCl, pH 8.3).
   - Since the active group can hydrolyze at the coupling pH, this step should be carried out in just a few minutes so the ligand can be applied as soon as possible

4. Transfer activated Sepharose to a flask containing 60 - 400 nmol ligand in 30 ml Coupling Buffer.
   - Be sure to set aside some of the ligand solution to determine binding efficiency

11

5. Mix gently, rocking or shaking for 2 hr at room temperature or overnight at 4°C (do not use a magnetic stirrer which can fragment the Sepharose beads).
   • Determine ligand binding efficiency (see Notes below)

6. Wash matrix with 200 ml Coupling Buffer on sintered glass funnel with gentle suction.

7. Transfer matrix back to flask, then incubate in 100 ml Blocking Buffer (1 M ethanolamine, adjusted to pH 8.0 with HCl) at room temperature for 2 hr or overnight at 4°C.

8. On sintered glass funnel, wash matrix with 100 ml Coupling Buffer, then 100 ml Acetate Buffer (0.1 M sodium acetate, 0.5 M NaCl, pH 4).

9. Wash with four cycles of 100 ml Coupling Buffer followed by 100 ml Acetate Buffer.

10. Finish the washes with Coupling Buffer containing 0.02% sodium azide before storage.

11. Matrix is now ready for packing into column.
   • Columns for affinity chromatography often are short and wide, with the matrix height being about twice the diameter.

Notes

- To estimate binding efficiency, determine the concentration of the ligand in solution before and after the coupling step (for protein ligands, see Chapter 3, Protein Concentration Determination or the discussion in Pepper [1992b]). Generally, 70 - 80% binding is optimal: lower binding leads to reduced column capacity while higher binding may result in reduced binding efficiency due to steric hindrance.

- A number of conditions can lead to poor coupling: low ligand concentration, suboptimal pH, impure ligand, improperly prepared matrix, inaccessibility of ligand (due to steric hindrance or large size: addition of a nonionic detergent or chaotropic agent may help), or improper buffer. Note that buffers such as Tris which contain an amino group should not be used or they may couple to the matrix.

- If the ligand is unstable at basic pH, coupling can be carried out in a lower pH buffer, but the efficiency is reduced.

11

B. Nonspecific Elution Strategies

Proteins bound to an affinity column can be eluted with saturating concentrations of a competing substrate or under other column conditions which reduce protein-ligand affinity. Substrate-mediated elution is the most specific and allows well-defined separation of the protein of interest without otherwise perturbing the column conditions, though subsequent removal of the substrate from the affinity column (often by extensive washing) or from the eluted protein (typically by dialysis) is required. Since specific affinity elutions are tailored to each situation, our discussion here will center on nonspecific protein elution methods; the specialized techniques section (Section III) will describe specific affinity eluents for each application. Purification on protein or peptide columns most closely resembles immunopurification, and readers are referred to the immunopurification protocol (Section III.A.2) for methods. Scopes (1994) provides a discussion of specific affinity elutions.

The mildest elution conditions are required if the protein of interest is labile. Optimal elution strategies require an understanding of the types of interactions between the protein and the ligand. To define the elution conditions, the ligand can be coated on wells of a 96-well ELISA plate, the protein solution added, and different elution solutions applied in different wells (Brahimi et al., 1993). The usual procedure when elution conditions have not been defined is to try the mildest elution conditions first and proceed to harsher treatments. Table 11.2 outlines possible elution conditions, arranged in approximate order from mildest to harshest.

Table 11.2
Nonspecific Elution Conditions for Affinity Purifications

To disrupt electrostatic interactions	Increase ionic strength

To disrupt hydrophobic interactions	Decrease ionic strength
	Include nonionic detergent
	Include polarity-reducing agent
	  (20% ethanol, up to 50% ethylene
	  glycol)

Temperature effect	Reduce temperature, usually
Low pH	0.1 M glycine-HCl (pH 2.5) 1 M propionic acid
High pH	0.15 M NH$_4$OH (pH 10.5) 0.1 M Na-CAPS (pH 10.7) 0.05 M triethanolamine (pH 11.5)
Chaotropic Agents	4 M MgCl$_2$ in 10 mM PBS (pH 7.0) 2.5 M NaI (pH 7.5) 3 M NaSCN (pH 7.4)*

Protein Deforming Agents
  (The concentrations listed are upper limits.
  Lower concentrations may be effective)
	8 M Urea (pH 7.0)*
	6 M Guanidine HCl*
	50% ethanediol (pH 11.5)
	10% dioxane in acetic acid
	50% ethylene glycol (pH 8 - 11.5)
	1% SDS*
	1% deoxycholate*

* These treatments denature most proteins

11

- To reduce contaminant binding to an affinity column, wash at high or low ionic strength to reduce nonspecific ionic or hydrophobic binding, respectively. Washing with a low concentration of the eluting agent is also often useful.

- As with ion exchange chromatography, gradient elutions may yield sharper peaks.

- Use of excess affinity support may allow the elution peak to broaden.

- Inclusion of dithiothreitol (DTT) or β-mercaptoethanol in any chromatography buffers can lead to broken disulfide bonds on proteins, possibly denaturing the protein of interest or disrupting disulfides linking protein complexes.

- Combinations of two eluents with different elution properties at lower concentrations may disrupt ligand-protein interactions and allow recovery of protein with higher levels of activity.

- Expected yield is 40 - 70% of starting material.

11

III. Specialized Techniques

A. Antibodies

1. Antibody Purification

a. Introduction

Since antibodies are increasingly important as reagents for affinity chromatography, a description of the most common method for purifying antibodies is in order. Antibodies are usually isolated from serum, hybridoma supernatant, or ascites fluid. While some immunological procedures can be carried out using these impure solutions, antibody enrichment is frequently required for satisfactory results, particularly if the antibodies are to be used for affinity chromatography. Purified antibodies can also be used for immunoassays, immunoblotting (see Chapter 8), enzyme conjugation, or cell staining (Harlow & Lane, 1988).

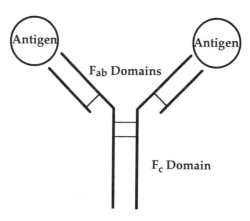

Figure 11.2. Antibody Structure

Although conventional protein purification methods (precipitation or chromatography techniques) can be applied to antibody purification, specialized procedures employing bacterial cell wall proteins offer a simple and highly specific separation. Protein A is purified from *Staphylococcus aureus* and Protein G from group G streptococci. These proteins specifically bind the Fc portion of immunoglobulins (Fig. 11.2). Since these proteins bind with different affinities depending on immunoglobulin type and specific species (Table 11.3), the proper purification reagent must be selected for a successful separation. For example, purification of polyclonal antibodies generated in a rabbit can be carried out using Protein A whereas separation of antibodies generated in a goat requires a Protein G column. Compatibilities of proteins A and G with antibodies from different species are compared in Table 11.3. Matrices are available with chimeric proteins possessing both Protein A and G binding sites.

The binding of antibodies to Protein A or Protein G can be altered by pH, salt concentration, and temperature. Protein A or G can be coupled to a matrix in the laboratory (see Section II above), but conjugated matrices are also available from commercial sources. Native Protein A and G also contain albumin binding sites which can lead to co-purification of contaminating proteins, but engineered variants are available from various suppliers which have had these sites eliminated. A protocol for running a Protein A or Protein G column is provided below.

Table 11.3
Binding Strength of Protein A or G to Immunoglobulins from Different Species

	Protein A	Protein G
Human		
IgG1	strong	strong
IgG2	strong	strong
IgG3	no binding	strong
IgG4	strong	strong
IgM	usually weak	no binding
Mouse		
IgG1	weak	strong
IgG2a	strong	strong
IgG2b	strong	strong
IgG3	strong	strong
IgM	weak	weak
Rat		
IgG1	weak	weak
IgG2a	no binding	strong
IgG2b	no binding	strong
IgG2c	strong	strong
Chicken	no binding	no binding
Rabbit	strong	strong
Goat		
IgG1	weak	strong
IgG2	strong	strong
Sheep		
IgG1	weak at best	strong
IgG2	strong	strong

11

Traditional purification procedures which can be used for purification of antibodies include precipitation methods (i.e., ammonium sulfate or caprylic acid), or chromatographic methods (ion exchange or gel filtration). However, these methods require multiple steps for purification (for example, ammonium sulfate precipitation followed by DEAE chromatography), so these techniques are recommended only if the affinity methods described below are not effective (for example, purification of IgM [Coligan et al., 1992]). One additional option is to use an anti-antibody immunoaffinity column comprised of an antibody specific for the Fc portion of the antibody of interest. Harlow & Lane (1988), Page et al. (1994) and Coligan et al. (1992) describe antibody purification protocols employing these procedures as well as protocols for antibody purification on ligand affinity columns.

It is important to test the antibody binding activity during the purification to be certain that the antibody of interest retains its activity.

Table 11.4
Typical Antibody Concentrations
(adapted from Harlow & Lane, 1988)

Source	Antibody Type	Antibody Concentration (mg/ml)
Serum	Polyclonal	10 mg/ml
Tissue Culture Supernatant, Serum-free Media	Monoclonal	0.0005 - 0.05 mg/ml
Ascites	Monoclonal	1 - 10 mg/ml

b. Protocol

Antibodies bound to Protein A or Protein G columns are most commonly eluted at low pH. Occasional limited antibody stability under acidic conditions makes it a wise practice to collect fractions in tubes containing a well-buffered, neutral solution. Other elution buffers are described in Section II.B. Typical antibody concentrations are listed in Table 11.4.

Preparation of Protein A- or Protein G-Sepharose Column

1) If starting with Protein A or G, couple to CNBr-activated Sepharose as described in Section II.A.

2) Hydrate Protein A-Sepharose (or Protein G-Sepharose) in starting buffer (100 mM Tris-HCl, pH 7.5, 100 mM NaCl) at room temperature for 30 min. Subsequent steps are often carried out at room temperature. 10 ml of buffer are used per 1 g of matrix, which will expand to 3 - 4 times its dry volume upon hydration.

3) Pack the matrix in a column and wash with 5 - 10 column volumes of starting buffer. Flow rates up to 1 ml/min are recommended.

4) Wash contaminants from column using 3 - 5 column volumes of 0.1M glycine-HCl (pH 2.5).

5) Reequilibrate column with 5 - 10 column volumes of starting buffer. If storing the column at this stage, include 0.02% NaN_3 and store at 4°C.

Preparation of Antibody Solution

6) Centrifuge sample solution at 10,000 x g for 10 min to remove precipitated material.

7) Dilute sample with an equal volume of starting buffer or add one-tenth volume 10x starting buffer to achieve proper pH and ionic strength.

Sample Application

8) Apply sample to column (up to 2 ml polyclonal antibody serum sample per ml of packed gel or up to 20 ml monoclonal antibody supernatant as long as total IgG is <80% of column capacity. Columns can bind 5 - 20 mg antibody per ml of wet beads).

9) Wash column with 10 column volumes starting buffer. Allow A_{280} to reach background levels before beginning elution.

11

Sample Elution

10) Elute with 5 - 10 bed volumes of 0.1 M glycine-HCl (pH 2.5), collecting 1 ml fractions in tubes containing 0.1 ml 1 M Tris-HCl (pH 8.0) to neutralize.

11) Pool peak fractions with A_{280} above 0.2.

12) To exchange buffer or remove salt, dialyze against phosphate-buffered saline (PBS: 137 mM NaCl, 2.7 mM KCl, 6.5 mM Na_2HPO_4, 1.5 mM KH_2PO_4), run a gel filtration column (for example a spin column, see Chapter 10), or use ultrafiltration (for example, Centricon, see Chapter 4).

13) Make 1 ml aliquots and store at 4°C for immediate use or -20°C for longer term storage.

14) Clean column with 5 column volumes 0.1 M glycine-HCl (pH 2.5) and reequilibrate with 10 column volumes starting buffer (containing 0.02% sodium azide for storage).

11

c. Discussion

- Properly maintained Protein A or Protein G columns can be utilized at least ten times with no loss of binding activity.

- **Antibody Titer**: The relative binding activity of antibody solutions is measured by the **titer**, that is, the dilution of the antibody solution which gives one-half the maximal signal with the detection system being used (such as immunoblotting or ELISA). Determining the titer (for example, by making sequential two-fold dilutions of the antibody solution) following a purification step provides a functional measure of the antibody concentration.

- **Antibody Storage**: Antibodies should be stored in concentrated form (at least 1 mg/ml) at a neutral pH. Long-term storage should be at -20°C or below. Antibodies are stable for years under these conditions, even with salt up to 150 mM. Repeated freeze-thaw cycles may inactivate some antibodies. In some cases, storage at 4°C may reduce antibody activity. Consult Harlow & Lane (1988) for more information about antibody storage.

- **Antibody Quantitation**: As a rule of thumb, 1 mg/ml of pure IgG has an absorbance at 280 nm of approximately 1.4. The investigator may also use standard protein concentration determination methods (Chapter 3) or SDS-polyacrylamide gel electrophoresis (Chapter 5). Unreduced SDS-PAGE (utilizing no β-mercaptoethanol) will yield a band of MW 150 kD, whereas SDS-PAGE containing β-mercaptoethanol should reveal two heavy chains at 50 kD and two poorly staining light chains at 25 kD.

- Although this method is commonly used, acid elution of antibodies from a Protein A or G column can denature some antibodies. Antigen binding activity of antibody solutions should be tested before and after purification. Coligan et al. (1992) recommend elution of mouse antibodies with 0.1M citric acid buffer: pH 6.5 for IgG1, pH 4.5 for IgG2a, and pH 3.0 for IgG2b or IgG3. For a description of other elution methods for affinity chromatography, see Section II.B.

- Binding characteristics of purified antibodies can be tested by enzyme-linked immunosorbent assay (ELISA), immunoblotting, or immunoprecipitation. See Harlow & Lane (1988) or Coligan et al. (1992).

- Polyclonal antibodies can be purified to be relatively monospecific with an antigen affinity column. For example, an antibody generated against a peptide can be purified by an affinity column employing that peptide.

- For protocols describing antibody labeling, see Harlow & Lane (1988) or Wisdom (1988).

11

- The generation of monoclonal or polyclonal antibodies is described in Coligan et al. (1992), Harlow & Lane (1988), Gullick (1988) and Amero et al. (1988).

2. Immunopurification

a. Introduction

Affinity purification using antibody columns is an extremely powerful method of protein purification, commonly permitting a 1000- to 10,000-fold purification in a single step. When the binding specificity of an antibody for the protein of interest has been defined and protein elution from the column does not irreversibly damage the protein, immunoaffinity purification is an excellent protein purification method, especially in the final stages of a purification procedure.

An immunoaffinity column requires that the antibody first be covalently attached to a matrix (described in Section II.A.). Alternatively, the antibody may be bound to Protein A or G beads and then immobilized with a cross-linking reagent (Harlow & Lane, 1988). This procedure allows proper antibody orientation for maximal access of the antibody binding site, since Protein A or G bind the Fc antibody region (Fig. 11.2). The antigen can then be applied to the column, the contaminants washed away, and the antigen eluted. Elution from an immunoaffinity column can be difficult if the nature of the protein binding is not understood. If a specific ligand is not available which can disrupt the antibody-antigen interaction, nonspecific conditions for disrupting protein interactions are required. Section II.B describes such nonspecific conditions, which must be applied with care lest the protein of interest be damaged. Other references for details about immunoaffinity chromatography include Kenney et al. (1988), Jack (1992), Harlow & Lane (1988), and Pharmacia (1993).

Monoclonal antibodies generally are more useful than polyclonal antibodies for immunoaffinity protein purification. Monoclonal antibodies represent a homogeneous antibody population with specificity for a single binding site on the protein of interest, thus allowing well-defined binding and more uniform protein elution since only one type of bond interaction must be disrupted for protein release. Nevertheless, antibodies from a polyclonal source can sometimes prove useful for affinity purification, especially polyclonals generated against a very pure antigen. Thus, the primary challenge in developing an effective immunopurification is to find a specific antibody with adequate binding affinity for the substrate (so the protein remains attached to the column during washing) which doesn't bind so tightly that extreme elution conditions are required (causing protein denaturation). A valuable approach to characterizing a useful antibody for immunopurification is to test a panel of monoclonal antibodies for those with adequate binding affinities which will in addition allow elution of an intact antigen (see Discussion below).

The quality of an immunopurification also depends on the purity of the antibody solution (see Section III.A. above). As a result, immunoaffinity columns are usually fairly expensive to make and their binding capacity can be relatively low (often less than 1 mg of antigen per ml of matrix). Therefore, small, fat columns are frequently prepared, and a purification protocol may call for an extract to be cycled over the column several times, allowing more antigen to be purified during each purification cycle. Smaller columns also limit the loss of valuable antibody in the event of column contamination or protease inactivation.

Immunoprecipitation (Chapter 4) can be considered another version of immunoaffinity purification. A variant of immunoaffinity chromatography is a 'negative immunopurification,' in which a specific contaminant is removed from a solution on an antibody column.

b. Protocol

The following protocol assumes that the antibody column is packed. A buffer pH that stabilizes the ligand should be used. Remember never to allow the matrix to run dry. All solutions should be filter-sterilized through a 0.22 µm filter to maximize column life. To determine the antigen binding capacity of the matrix, a trial binding experiment can be conducted in test tubes by mixing increasing amounts of antigen with a constant amount of antibody bound to matrix. The following sample protocol employs elution at low pH, presupposing that the antigen is stable under these conditions.

Column Equilibration
1) Wash column with 5 - 10 column volumes of starting buffer (for example, PBS).
2) Wash contaminants from column with 3 - 5 column volumes of elution buffer (for example, 0.1 M glycine-HCl, pH 2.5).
3) Equilibrate column with 5 - 10 column volumes of starting buffer.

Sample Loading
4) Equilibrate sample in starting buffer (by dialysis, gel filtration, or by adding 1/10 volume 10x starting buffer to sample). Sample should be clarified previously by centrifugation (100,000 x g for 30 min) or filtration (0.22 µm pore size filter).
5) Apply sample to column. Slower flow rates allow for better antigen binding; sample may also be reapplied several times.

Column Washing

6) Wash with 5 - 10 column volumes of starting buffer or until A_{280} approaches background levels. If the antibody - antigen interaction is strong, include salt in the starting buffer (for example 0.5 M KCl) to reduce nonspecific binding.

Protein Elution

7) Apply 2 column volumes of elution buffer (for example, 0.1 M glycine-HCl, pH 2.5), collecting 1 ml fractions in tubes containing 0.1 ml 1 M Tris-HCl (pH 8.0) to neutralize the pH.
8) Assay fractions and combine active fractions.
9) Exchange buffer (immediately if the antigen has limited stability) by dialysis or gel filtration as necessary.

Regeneration of column

10) Wash with 10 volumes 0.2 M glycine/HCl (pH 2.5), then with 10 - 20 volumes PBS (containing 0.02% azide if for storage)

c. Discussion

- Although high-affinity antibody binding is desirable (especially for antigen purification from dilute solutions), overly strong binding can prevent elution. An intermediate binding affinity (10^{-7} M) is often best (Jack, 1992). Pepper (1992a) provides recommendations for antibody screening.
- The slightest microbial contamination of an immunoaffinity column can result in antibody proteolysis, so use of sterile buffers and extracts is essential, as is adherence to careful column storage conditions.
- Immunoaffinity chromatography can often be carried out at room temperature.
- To reduce nonspecific background binding, the antigen solution can be passed over a column of uncoupled matrix.

11

B. Nucleic Acid Affinity Chromatography

1. Introduction

Nucleic acid affinity columns have greatly enhanced the characterization of regulatory proteins involved in the control of gene expression, chromosome repair and replication, and genetic recombination. Nucleic acid binding proteins may bind single-stranded DNA, double-stranded DNA, or RNA. DNA binding proteins can be sequence-specific or may bind DNA nonspecifically. In addition, affinity resins containing specific nucleotides are valuable for isolation of enzymes which utilize structurally similar cofactors such as NAD or NADP. In sum, nucleic acid affinity chromatography is a valuable tool in the purification and characterization of a wide range of proteins.

Our discussion will focus on the use of an oligodeoxynucleotide affinity matrix for protein purification, stressing primarily the technical problems which are unique to nucleic acid affinity chromatography. Other approaches can be readily applied using the information provided in this section, including, for example, purification with biotinylated DNA coupled to magnetic beads.

Before embarking on an affinity purification using oligonucleotides, some fundamental requirements need to be defined. Most importantly, a minimum nucleic acid binding sequence must be determined (see Discussion for description of typical binding assays). Kadonaga (1991) notes that successful affinity purifications have used oligonucleotides containing between 14 and 51 bases. Secondly, the length of the spacer separating the recognition sequence from the matrix can affect protein accessibility. Often a longer oligonucleotide can be used instead of a spacer. Finally, the matrix used for immobilization should have adequate flow properties while permitting access of a protein or protein complex to the nucleotide sequence. Most applications employ Sepharose 4B, and the protocol below will describe the use of CNBr-activated

Sepharose 4B for ligand immobilization and affinity chromatography.

An important consideration in nucleic acid affinity chromatography is nonspecific binding. Since the affinity ligand often comprises only four nucleotides, purification of nonspecific proteins is frequently a problem. Three basic approaches are recommended. The first approach exploits a nonspecific oligonucleotide affinity precolumn. This column can contain an underivatized matrix (such as Sepharose 4B) or an affinity column containing a random or scrambled oligonucleotide sequence. A second approach is to include a large excess of soluble nonspecific oligonucleotide in the affinity column to prevent column binding of nonspecific nucleic acid binding proteins. The third approach is recommended if nonspecific proteins are binding to sequences flanking the high affinity binding site. In this case, a second affinity column can be constructed, substituting an alternative flanking sequence. The use of either a nonspecific affinity precolumn, soluble nonspecific DNA in the affinity column, or an additional affinity column employing alternative flanking sequences usually solves nonspecific protein binding problems.

11

2. Protocol

Coupling an Oligonucleotide to CNBr-activated Sepharose 4B

The following protocol describes the coupling of an oligonucleotide to cyanogen bromide-activated Sepharose 4B. DNA coupling to CNBr-activated Sepharose occurs through the bases themselves. Although in principle this method of attachment might interfere with accessibility of the desired binding sequence, this simple procedure has been widely successful. For descriptions of how to prepare and purify oligonucleotides, consult Kadonaga (1991) or Hornby et al. (1992). Quantitation of incorporation efficiency can be achieved by measuring the absorbance of the matrix at 260 nm or, more accurately, by hydrolyzing the nucleic acid from the support and conducting a phosphate determination (Arndt-Jovin et al., 1975).

a) Suspend 1.5 g dry CNBr-Sepharose 4B in water (should make about 5 ml of swelled gel). Filter through a sintered-glass funnel, stopping the vacuum while the Sepharose is still moist.

b) On the funnel, wash with 200 ml 1 mM HCl followed by 200 ml water and 200 ml 10 mM potassium phosphate (pH 8.0). All solutions should be at 4°C.

c) Recover Sepharose in a 15 ml tube containing 2 ml 10 mM potassium phosphate (pH 8.0).

d) To the activated CNBr-Sepharose, add 50 µl oligonucleotide (containing about 2 nmol nucleotide per ml of gel) and shake at room temperature for 16 h.

e) Transfer to a sintered-glass funnel and remove liquid.

f) Wash Sepharose with 200 ml water, then 100 ml 1M ethanolamine-HCl (pH 8.0)

g) Resuspend resin in 7 ml 1 M ethanolamine-HCl (pH 8.0), shake in a polypropylene tube at room temperature for 4 - 6 h. This step inactivates uncoupled CNBr-activated Sepharose.

h) Wash Sepharose with 100 ml 10 mM potassium phosphate (pH 8.0), then 100 ml 1 M potassium phosphate (pH 8.0), then 100 ml 1 M KCl, then 100 ml water, then 100 ml storage buffer (10 mM Tris-HCl [pH 7.6], 0.3 M NaCl, 1 mM EDTA, 0.02% NaN$_3$)

i) Store at 4°C for up to a year.

Nucleic Acid Affinity Chromatography

Before applying a protein solution to a nucleic acid affinity column, it is recommended that an extract first be enriched for the protein of interest by another purification method (such as ammonium sulfate precipitation [Chapter 4], ion exchange chromatography [Chapter 9], or gel filtration chromatography [Chapter 10]). The additional purification step will remove most of the contaminants and decrease the likelihood of nonspecific binding. Affinity chromatography columns are generally wide and short (for example, a length-to-width ratio of 2:1).

a) Equilibrate the column with 10 column volumes of affinity chromatography buffer (AC Buffer: for example, 20 mM Tris-HCl [pH 8.0], 0.15 M KCl, 1 mM EDTA).

b) Adjust protein sample to similar ionic concentration (0.15 M KCl).

c) Add nonspecific DNA to protein sample at 100 ng/μg protein and incubate on ice for 10 min. Nonspecific DNA is often 1 kilobase in length and can be double stranded poly(dI-dC) or poly(dA-dT), or sonicated DNA from *E. coli*, calf thymus or salmon sperm. To remove complexes which may be formed with this treatment, a 10 min centrifugation at 10,000 x *g* is recommended.

d) Slowly apply protein solution to affinity column (15 ml/hr).

e) Wash matrix with 5 column volumes AC buffer

f) Elute proteins with 1 column volume each of AC buffer containing 0.2, 0.3, 0.6, 1.0 M KCl.

11

g) Test fractions for activity and pool fractions with significant activity. Since protein concentrations are often low, the nonionic detergent Triton X-100 (0.05%) is often added to reduce protein adsorption to the container wall (though this detergent interferes with absorbance measurements at 280 nm).

h) Store active protein fractions at -70°C.

i) Regenerate affinity column by washing with 10 column volumes of AC buffer containing 2.5 M NaCl and 0.5 M EDTA, then reequilibrate and store the column with AC buffer containing 0.02% NaN_3.

11

3. Discussion

- Specific elution of nucleic acid binding proteins using the binding sequence can be effective. Apply a high concentration of the oligonucleotide to the column for elution.

- Common assay methods for DNA binding proteins are filter-binding (Ausubel et al., 1994), gel retardation (Fried & Crothers, 1981), and DNAse footprinting (Galas & Schmitz, 1978). Note that the activity assay should be used to define the amount of nonspecific competing DNA to be used during chromatography.

- DNA removal from the protein solution may enhance yields by removing soluble specific DNA sequences. DNA may be digested with DNAse or precipitated with 1% polyethyleneimine or streptomycin sulfate (up to 5%). After DNAse treatment to remove soluble oligonucleotides, buffers should contain EDTA to remove divalent cations and inactivate the DNAse.

- DNA coupling to the matrix can be calculated by determining the solution absorbance at 260 nm before and after coupling.

- Some variables which can affect DNA-protein binding and elution include temperature, pH, and magnesium concentration.

- Protein binding to DNA may require cofactors (ATP, Mg, S-adenosylmethionine). Thus, cofactor removal may allow elution (Alberts and Herrick, 1971)

11

- CNBr-activated matrices will allow slow oligonucleotide displacement, so columns should be discarded after 5 - 10 chromatography runs.

- Since most DNA binding proteins are basic, a common approach is to precede an affinity column purification step with ion exchange chromatography. Nucleic acid binding proteins can be further purified with some nonspecific chromatographic methods, including heparin, phosphocellulose, and lectin chromatography.

- An extensive list of original references employing DNA affinity chromatography can be found in Kadonaga (1991). In addition, Jarrett (1993) provides a useful review of nucleic acid affinity chromatography.

11

C. Lectin

1. Introduction

Lectins are proteins that bind carbohydrates reversibly. Since most lectins contain at least two carbohydrate binding sites per molecule, they can precipitate glycoproteins and agglutinate cells. Lectins are also useful as affinity ligands for glycoproteins, with simple sugars permitting mild protein elution conditions. Although lectin affinity chromatography is not highly selective, lectin columns have nonetheless become valuable tools in the arsenal of the protein purifier, usually as one in a series of purification steps. A lectin column can also be used for the removal of a glycoprotein contaminant. Table 11.5 shows a list of some commonly used lectins, many of which are commercially available conjugated to a matrix.

Lectins should be chosen for affinity chromatography according to their specificity and tightness of binding. For example, Concanavalin A (Con A) binds glucose- or mannose-containing proteins whereas wheat germ lectin binds proteins with N-acetylglucosamine (for a more detailed description, see Lotan & Nicolson, 1979). Membrane glycoproteins often bind wheat germ agglutinin while soluble glycoproteins can usually be purified with Con A or lentil lectin columns. A discussion of oligosaccharide structure and its importance in lectin binding can be found in Sutton (1989). Since in many cases the exact composition of the glycoprotein to be purified is not known, it is frequently useful to test a variety of lectins for binding to the protein of interest. Kits of lectins are available for this purpose (Vector Labs).

11

Some general considerations which are unique to lectin affinity chromatography are important to keep in mind. As listed in Table 11.5, some lectins require certain divalent cations for their structural integrity. Since lectin affinity chromatography frequently involves membrane proteins, detergents are often required to keep these proteins soluble. Although lectins are rather tolerant of nonionic detergents, ionic detergents can reduce binding significantly.

Table 11.5
Lectins

Lectin	Specificity	Eluting Saccharides	Protective Saccharide	Ion Requirement
Concanavalin A (ConA)	α-D-mannose, α-D-glucose	α-D-mannose, α-D-glucose, methyl-α-D-mannoside	methyl-α-D-mannoside	Ca^{2+}, Mn^{2+}
Lentil Lectin (Lens culinaris)	α-D-mannose, α-D-glucose	α-D-mannose, α-D-glucose, methyl-α-D-mannoside	methyl-α-D-mannoside	Ca^{2+}, Mn^{2+}
Wheat Germ Agglutinin	N-acetyl glucosamine	N-acetyl glucosamine, sialic acid	N-acetyl glucosamine	Ca^{2+}, Mn^{2+}, Zn^{2+}
Jacalin	α-D-galactose	D-galactose, melibiose	D-galactose	none
Ricinus communis	α-D-galacto-pyranose	N-acetyl galactosamine, lactose, D-galactose	methyl-β-galactoside	none

2. Protocol

Lectin affinity chromatography is relatively easy to carry out. The lectin can be immobilized on the matrix by a variety of methods, with cyanogen bromide coupling being relatively popular. Coupling of a lectin to CNBr-activated Sepharose can be carried out as described in section II.A., substituting 1 - 10 mg of lectin per ml of matrix material. It is critical to include the appropriate divalent cation (0.1 mM) and protective sugar (5%, w/v) (see Table 11.5 and Sutton, 1989) so that the binding site is inaccessible during the coupling procedure. Many lectin affinity columns are also available commercially.

The following protocol describes the purification of a protein on a Con A affinity column. Procedures for other lectin columns should be similar, substituting the appropriate eluting sugar. Chromatography columns which are relatively long and narrow provide the best resolution whereas short, fat column allow faster flow rates with proteins eluting in a smaller volume. Small trial columns (for example, packed in a Pasteur pipet) or batch chromatography tests in Eppendorf tubes can be used to estimate column capacity.

a) Prepare and pack the Con A column. Refer to section II.A. above for lectin immobilization instructions and to Chapter 9, section I.B.3. for a description of column packing.
b) Equilibrate the column by washing with 10 bed volumes of chromatography buffer (20 mM Tris-Cl, pH 8.0, 0.15 M NaCl, 1 mM $MnCl_2$, 1 mM $CaCl_2$, 30 mM octyl glucoside)
c) Prepare glycoprotein by dialyzing into chromatography buffer or diluting 1:1 with chromatography buffer. Centrifuge or filter the protein solution to remove precipitate.
d) Apply protein to column.
e) Wash with chromatography buffer until absorbance at 280 nm returns to baseline.
f) Elute with five bed volumes 10 mM methyl-α-D-mannoside in chromatography buffer

g) Continue step elutions with five bed volumes of 20 mM, 50 mM, 100 mM, 250 mM, and 500 mM methyl-α-D-mannoside in chromatography buffer.

h) Analyze fractions for protein of interest. Proteins can be dialyzed to remove eluting sugar.

i) Regenerate column with chromatography buffer containing 1 M NaCl.

j) Column can be stored in chromatography buffer containing 0.02% sodium azide to prevent against column contamination.

3. Discussion

- To improve lectin-glycoprotein binding, slow the flow rate when applying the protein solution to the column. Alternatively, a batch chromatography step can be performed to allow maximal protein-lectin interaction (see Chapter 9, Section III). In addition, increasing the column temperature, up to 37°C, may improve protein binding.
- Nonspecific binding can be reduced by preceding the elution step with a high salt wash.
- Several modifications can be tried to improve protein elution:
  - Stop the column flow after applying the eluent to increase the time for protein-lectin dissociation. An overnight incubation may improve protein recovery.
  - In situations where warming the column has improved protein binding, cooling will often exert the opposite effect, promoting elution.
  - Borate buffers (0.1 M, pH 6.5) can help with a difficult elution since borate complexes some sugars.

- CNBr-activated matrices may contain charged groups which can introduce an ion exchange effect. This effect is mostly eliminated by including 0.1 M salt in the chromatography buffer.
- Some detergents which have been used successfully for lectin affinity chromatography include Triton X-100, Nonidet P-40, and Octyl β-D-glucoside. Note that TX-100 and NP-40 absorb strongly at 280 nm.
- If membrane proteins are to be purified, the lectin should be coupled to the matrix via a hydrophilic rather than a hydrophobic spacer arm to minimize nonspecific hydrophobic interactions. In these cases, ethylene glycol (up to 50%) can help to reduce hydrophobic interactions. A general overview of membrane protein purification can be found in Thomas & McNamee (1990).
- Step elutions have been generally more successful than gradient elutions.
- Avoid the use of phosphate buffers which precipitate many divalent cations (see Chapter 1).
- Some methods for detection of glycoproteins are described in Gerard (1990).
- In addition to CNBr-activated Sepharose, other activated matrices include AffiGel 10 and AffiGel 15 (Bio-Rad) and carbonyldiimidazole (CDI) agarose (West & Goldring, 1992).

11

D. Dye Ligand Chromatography

1. Introduction

Dye ligand chromatography is not truly affinity chromatography in that dyes are not the natural ligands of the proteins that bind them. Nevertheless, dye columns can bind proteins remarkably well and lead to significant purifications. Indeed, binding can sometimes be tighter than that achieved by a normal ligand. Dye ligand columns are generally inexpensive, stable, and display high capacities for protein binding. Thus, dye ligand chromatography can serve as a valuable step in a protein purification protocol.

Selecting an appropriate dye for purifying a specific protein is generally by trial and error. Cibacron Blue F3GA, the 'preeminent' dye in the field, resembles nicotinamide adenine dinucleotide (NAD) and has been generally useful in purifying kinases, hydrolases, polymerases and other nucleotide-dependent proteins. However, this binding specificity is not universal and, in addition, Cibacron Blue has been used in the purification of various proteins with no nucleotide binding function. Protein binding to dyes can involve hydrophobic, electrostatic, or hydrogen bonding forces. Therefore, no standard method is available for selecting the best dye for chromatography of a given protein, although kits are available commercially for screening a variety of dyes. Scopes (1992) groups dyes according to protein binding ability and recommends a system for selecting the most effective dye.

Dye ligand chromatography is relatively simple to perform. Matrices are commercially available (Table 11.6) on a number of support resins and these are recommended for better reproducibility. An extremely useful approach is to run sequential dye ligand columns, the first of which doesn't bind to the protein of interest while the second does. Most protein elutions are then accomplished at high salt, though other approaches are mentioned below. Poor protein binding can often be improved by lowering the solution pH or by including divalent or trivalent cations. By following these few guidelines, dye binding chromatography can provide an important contribution to a protein purification protocol.

Table 11.6
Common Dyes Used in Dye Ligand Chromatography

Cibacron Blue F3GA
Procion Blue H-B
Procion Red HE 3B
Procion Yellow H-A
Procion Green H-E4BD

2. Protocol

Before embarking on a protein purification using dye ligand chromatography, an appropriate dye must be identified. Small scale experiments with different media (1 ml columns) are appropriate for screening of dyes. Scopes (1992) recommends a strategy for selecting media from groups of dyes with generally different binding characteristics. Further small scale experiments should allow optimization of elution parameters. A dye binding column which binds many proteins but not the protein of interest is an excellent step to include immediately prior to application of the protein solution to this column. Below is a Cibacron Blue F3GA dye binding protocol employing high salt elution.

11

a) Pack a column with 5 ml of agarose - Cibacron Blue F3GA. For most applications, a short, wide column should work well. Longer and narrower columns are more useful for proteins with weak binding characteristics, though flow rates will be reduced.

b) Equilibrate the column with 10 bed volumes of chromatography buffer (for example, 20 mM Tris-HCl, pH 8.0, 0.1 M NaCl).

c) Prepare the protein solution for loading by exchanging into chromatography buffer (by dialysis or by diluting 1:1 into chromatography buffer). The protein solution should be clarified by centrifugation or filtration prior to loading. 50 to 200 mg of protein can be loaded onto a 5 ml column. Unless the protein binding affinity is very high, the protein concentration upon loading should be relatively high (10 to 20 mg/ml).

d) Load protein onto column.

e) Wash with chromatography buffer for five bed volumes or until protein absorbance at 280 nm returns to baseline.

f) Elute protein with five bed volumes elution buffer (for example, 20 mM Tris-HCl, pH 8.0, 1.0 M NaCl).

g) Analyze fractions for protein of interest and pool active fractions.

h) Regenerate column with 3 bed volumes of 1 M NaOH, then wash with 10 bed volumes of chromatography buffer containing 0.02% NaN_3.

Alternative elution strategies

- A more specific elution would employ a ligand (such as NAD or NADP, at concentrations between 1 μM to 25 mM), if the dye occupies the ligand binding site.

- Removing a metal ion, with EDTA, can sometimes effect an elution.

- Increasing the pH can be more effective than a salt elution. More alkaline solutions tend to reduce protein binding to dye ligand columns.

- When hydrophobic interactions are important for protein binding, reducing the polarity of the solution (either with 20% ethanol, 10% isopropanol, or 10 - 50% ethylene glycol) should aid in protein elution, though peaks will tend to be broad.

- Chaotropic ions (such as urea, LiBr, or KSCN) can be used for elution, but they often denature proteins

- Nonspecific interactions can be reduced with the use of phosphate buffers, although this will result in precipitation of multivalent cations.

11

3. Discussion

- The most useful variables in effecting an elution are pH, ionic strength, and temperature. Higher pH or ionic strength tend to reduce protein binding.

- More stringent regeneration conditions include 6 - 8 M urea or guanidine HCl or 2 - 4 M sodium thiocyanate.

- A variety of matrices for dye ligand chromatography with excellent rigidity and flow characteristics are now available, including Sepharose, Ultrogel, Trisacryl, and Bio-Gel A.

- One must be careful, given the wide range of similarly colored dyes and different dye suppliers, to utilize the same dye column for a series of experiments. Different dyes and even different lots of the same dye have displayed altered chromatographic characteristics. Even dye ligand matrices from different manufacturers can have differing amounts of dye immobilized to the matrix.

- Lowe & Pearson (1984), Angal (1988), and Scopes (1992) provide protocols for immobilizing a dye to a matrix.

11

E. Immobilized Metal Affinity Chromatography

1. Introduction

Immobilized metal affinity chromatography (IMAC) takes advantage of interactions between exposed residues of proteins and metal ions bound on a matrix (Sulkowski, 1985). Surface amino acids which can act as electron donors (especially histidine) chelate a metal ion, thus preferentially detaining that protein on an IMAC column. Since the electron donor must be at least partially unprotonated to chelate the metal ion, more alkaline solutions favor stronger protein binding to an IMAC column.

In order to interact with a protein, the immobilized metal ion must be accessible. Accessibility is achieved by metal complexation at the end of a hydrophilic spacer arm extending from the matrix. Typically, commercially supplied iminodiacetic acid (IDA) Sepharose is used for metal chelation, although Tris (carboxymethyl) ethylenediamine (TED) Sepharose is also available when tighter matrix-metal binding is desired. Copper is a good metal to try first for IMAC when little information about a protein is available. Other metals which are often used include nickel, zinc, cobalt, and calcium. No good guidelines are available for selecting the best metal ion for purification with IMAC; only trial and error can be recommended. High salt (0.5 - 1.0 M NaCl) should be included during chromatography to reduce nonspecific ionic interactions. In addition, metal chelators (EDTA, EGTA, citrate) must be avoided prior to and during protein binding.

11

2. Protocol

Charging the Matrix

a) Pack a column with 10 ml IDA-Sepharose 6B. Column dimensions can vary according to how tightly the protein of interest binds; a tightly bound protein will still be well separated using a short, wide column.

b) Wash matrix with 3 bed volumes water.

c) Charge matrix with 3 bed volumes 1 mg/ml $CuSO_4 \cdot 5H_2O$.

d) Wash column with 5 bed volumes chromatography buffer (20 mM sodium phosphate, pH 7.5, 0.5 M NaCl).

Loading the Protein

e) Prepare protein solution by equilibrating with chromatography buffer (use dialysis or dilute 1:1 with chromatography buffer), then clarify by centrifugation or filtration. Matrix capacity should determine how much protein can be loaded safely: trial experiments are useful for determining column capacity. A reasonable estimate of capacity is 10 - 100 mg protein per ml of matrix.

f) Load protein solution (1 - 10 mg/ml) onto IMAC column.

g) Wash column with 5 bed volumes chromatography buffer or until absorbance at 280 nm returns to baseline.

Eluting the Protein and Regenerating the Matrix

h) Elute the protein of interest by washing column with 5 bed
    volumes elution buffer (100 mM sodium acetate, pH 5.0,
    0.5 M NaCl).

i) More strongly binding proteins can be eluted and the
    column regenerated with 50 mM EDTA in chromatography
    buffer.

•   Another elution strategy is to compete for the metal
    binding, for example, by increasing the concentration of
    imidazole buffer (Yip & Hutchens, 1992), histidine,
    glycine, or ammonium chloride. Chaotropic ions (urea) or
    detergents may also be useful for strongly binding proteins.
•   A gradient elution can be tried instead of the step elution
    described.

3. Discussion

•   Varying amounts of metal ion can achieve different
    purification results. Generally, charging the matrix with
    one-third to one-half of the metal binding capacity is best.
    Metal leaching is then minimized since additional matrix
    binding sites are available.
•   When eluting by lowering the pH, be sure not to inactivate
    the protein of interest nor to cause isoelectric precipitation.
    In addition, IDA-Zn^{++} and IDA-Co^{++} matrices become
    unstable upon lowering the pH (Sulkowski, 1985).
•   One strategy for quantitating protein elution is to include
    trace amounts of labeled metal ions during charging and
    monitor the eluted fractions for radioactivity.

11

F. Hydrophobic Interaction Chromatography

1. Introduction

Hydrophobic interaction chromatography (HIC) depends on an affinity between a hydrophobic group on the matrix and regions of hydrophobicity on a protein. Hydrophobicity stems from a repulsion of a nonpolar substance immersed in a polar liquid such as water (for more discussion of hydrophobic interactions, see Roe, 1989 and Kennedy, 1990). Membrane proteins of necessity possess significant hydrophobic regions for membrane anchoring. Soluble proteins can exhibit hydrophobic patches on their external surface (which may serve to facilitate protein complex formation) or may contain a hydrophobic ligand binding site or active site. Such exposed hydrophobic regions are ideal for separation by HIC. Although HIC exploits rather nonspecific affinities for protein separation and thus does not provide high resolution, this method is very useful since the principles involved are significantly different from those of other chromatographic techniques.

Although in principle HIC supports can contain alkyl or aryl chains of any size, in practice most separations employ phenyl or octyl groups (Fig. 11.3). As the hydrocarbon chain length increases (and becomes more hydrophobic), fewer, but more hydrophobic, proteins will adsorb. However, if the hydrophobic interaction is too strong, extreme elution conditions will be required and may lead to protein denaturation. Phenyl Sepharose, which is less hydrophobic than Octyl Sepharose, is often a good matrix to try at the outset of a hydrophobic purification step.

Phenyl Sepharose:	$- O - CH_2 - CHOH - CH_2 - O - C_6H_5$
Octyl Sepharose:	$- O - CH_2 - CHOH - CH_2 - O - (CH_2)_7 - CH_3$

Figure 11.3. Hydrophobic Interaction Matrices

Hydrophobic interactions become stronger as the solution salt concentration increases.   Not surprisingly, most HIC protocols call for protein loading at high salt and elution by lowering the salt concentration. Thus, HIC can be conveniently run after an ammonium sulfate precipitation or ion exchange chromatography. These mild elution conditions together with a high protein capacity (10 - 100 mg/ml) make HIC a valued step in many protein purifications and an alternative for buffer exchange.

2. Protocol

This protocol assumes that the protein of interest has just been subjected to ammonium sulfate precipitation at 75% ammonium sulfate and is soluble at 50% ammonium sulfate.

a) Pack 5 ml Phenyl Sepharose CL-4B into column.

b) Wash column with 10 column volumes Loading Buffer (20 mM sodium phosphate, pH 7.0, 50% ammonium sulfate).

c) Adjust protein sample to 50% ammonium sulfate in pH 7.0 sodium phosphate buffer, and apply 200 - 500 mg of sample to column.

d) Wash column with 3 column volumes Loading Buffer or until absorbance returns to background.

e) Elute protein with step gradient of 2 column volumes each pH 7.0 sodium phosphate buffer containing 40%, 30%, 20%, 10%, and 0% ammonium sulfate.

f) Regenerate column with 5 column volumes water, then 5 column volumes 1 M NaCl, then 5 column volumes water.

11

3. Discussion

- Proteins which bind weakly to Phenyl Sepharose may bind more tightly to Octyl Sepharose.
- Specific elutions might include cofactors, substrates, inhibitors, or metal ions.
- Alternative elution conditions can exploit increasing concentrations of polarity-reducing agents (for example, ethylene glycol, from 30% to 75%, or ethanol), chaotropic ions, nonionic detergents (such as Tween 20 or Triton X-100), or denaturants (1 M urea). In addition, increasing the pH or lowering the temperature may reduce binding. Note that polarity-reducing agents tend to denature proteins.
- A protein possesses its maximum hydrophobicity near its isoelectric point. Thus, hydrophobic binding can be decreased as the solution pH separates from the protein's pI.
- Shaltiel (1984) provides a discussion of elution strategies.
- A more effective elution may be to use two elution techniques simultaneously (for example, lower the salt concentration while increasing the concentration of ethylene glycol).
- Gradient elutions may improve resolution.
- Irreversible protein binding can occur if the matrix is too hydrophobic. A small scale trial can help to avoid excess protein losses.
- Since hydrophobic interactions are relatively nonspecific, it is possible for a protein to adsorb to the matrix at several sites, resulting in an irregular elution.
- Roe (1989) and Kennedy (1990) provide suggestions for column regeneration when using detergents for elution.
- Hydrophobic interaction columns can generally be used 20 to 30 times.

IV. References

Alberts, B. and G. Herrick. 1971. DNA-cellulose Chromatography. Meth. Enzymol. 21: 198-217.

Amero, S.A., James, T.C., and S.C.R. Elgin. 1988. Production of Antibodies Using Proteins in Gel Bands. pp. 355-362 in New Protein Techniques, J.M. Walker, ed. Humana Press, Clifton, New Jersey. 1988.

Angal, S. Dye-Ligand Chromatography. pp. 111-122 in New Protein Techniques, J.M. Walker, ed. Humana Press, Clifton, New Jersey.

Arndt-Jovin, D.J., Jovin. T.M., Baehr, W., Frischauf, A.-M., and M. Marquardt. 1975. Covalent Attachment of DNA to Agarose: Improved Synthesis and Use in Affinity Chromatography. Eur. J. Biochem. 54: 411-418.

Ausubel, F.M. et al. (eds.). DNA - protein interactions. Section 12. Current Protocols in Molecular Biology. John Wiley & Sons, New York. 1994.

Brahimi, K., Perignon, J.-L., Bossus, M., Gras, H., Tartar, A., and P. Druihle. 1993. Fast Immunopurification of Small Amounts of Specific Antibodies on Peptides Bound to ELISA Plates. J. Imm. Meth. 162: 69-75.

Coligan, J.E., Kruisbeek, A.M., Margulies, D.H., Shevach, E.M., and W. Strober, (eds.). 1992. Current Protocols in Immunology. John Wiley & Sons, New York.

Fried, M., and D.M. Crothers. 1981. Equilibria and Kinetics of Lac Repressor-operator Interactions by Polyacrylamide Gel Electrophoresis. Nucl. Acids Res. 9: 6505-6523.

Galas, D. J., and A. Schmitz. 1978. DNAase Footprinting: A Simple Method for the Detection of Protein-DNA Binding Specificity. Nucl. Acids Res. 5: 3157-3170.

Gerard, C. 1990. Purification of Glycoproteins. Meth. Enzymol. 182: 529-539.

Gullick, W.J. 1988. Production of Antisera to Synthetic Peptides. pp. 341-354 in New Protein Techniques, J.M. Walker, ed. Humana Press, Clifton, New Jersey.

11

Harlow, E., and D. Lane. 1988. Antibodies: A Laboratory Manual. 726 pp. Cold Spring Harbor Laboratory, Cold Spring Harbor, New York.

Hornby, D., Ford, K., and P. Shore. 1992. Purification of DNA Binding Proteins by Affinity and Ion Exchange Chromatography. pp. 273-285 in Practical Protein Chromatography, A. Kenney & S. Fowell, eds. Meth. Mol. Biol. Vol. 11.

Jack, G.W. 1992. Immunoaffinity Chromatography. pp. 125-133 in Practical Protein Chromatography, A. Kenney & S. Fowell, eds. Meth. Mol. Biol. Vol. 11.

Jarrett, H.W. 1993. Affinity Chromatography with Nucleic Acid Polymers. J. Chrom. 618: 315-339.

Kadonaga, J.T. 1991. Purification of Sequence-specific Binding Proteins by DNA Affinity Chromatography. Meth. Enzymol. 208: 10-23.

Kennedy, R.M. 1990. Hydrophobic Chromatography. Meth. Enzymol. 182: 339-343.

Kenney, A., Goulding, L., and C. Hill, C. 1988. The Design, Preparation, and Use of Immunopurification Reagents. pp. 99-110 in New Protein Techniques, J.M. Walker, ed. Humana Press, Clifton, New Jersey.

Lotan, R., and G.L. Nicolson. 1979. Purification of Cell Membrane Glycoproteins by Lectin Affinity Chromatography. Biochim. Biophys. Acta 559: 329-376.

Lowe, C.R., and J.C. Pearson. 1984. Affinity Chromatography on Immobilized Dyes. Meth. Enzymol. 104: 97-113.

Page, M., Baines, M.G., and R. Thorpe, R. 1994. Preparation of Purified Immunoglobulin G (IgG). pp. 407-432 in Basic Protein and Peptide Protocols, J.M. Walker, ed. Meth. Mol. Biol., Vol. 32. Humana Press, Totowa, New Jersey.

Pepper, D.S. 1992a. Selection of Antibodies for Immunoaffinity Chromatography. pp. 135-171 in Practical Protein Chromatography, A. Kenney & S. Fowell, eds. Meth. Mol. Biol. Vol. 11.

Pepper, D.S. 1992b. Some Alternative Coupling Chemistries for Affinity Chromatography. pp.173-196 in Practical Protein Chromatography, A. Kenney & S. Fowell, eds. Meth. Mol. Biol. Vol. 11.

11

Pharmacia. 1993. Affinity Chromatography: Principles and Methods. 143 pages.

Roe, S. 1989. Separation Based on Structure. pp. 175-244 in Protein Purification Methods: A Practical Approach, E.L.V. Harris & S. Angal, eds. IRL Press, Oxford.

Scopes, R.K. 1994. Protein Purification: Principles and Practice. Third Edition. Springer-Verlag, New York. 380 pages.

Scopes, R.K. 1992. Biospecific Affinity Elution. pp. 209-221 in Practical Protein Chromatography, A. Kenney & S. Fowell, eds. Meth. Mol. Biol. Vol. 11.

Shaltiel, S. 1984. Hydrophobic Chromatography. Meth. Enzymol. 104: 69-96.

Sulkowski, E. 1985. Purification of Proteins by IMAC. TIBTech. 3: 1-7.

Sutton, C. 1989. Lectin Affinity Chromatography. pp. 268-282 in Protein Purification Methods: A Practical Approach, E.L.V. Harris & S. Angal, eds. IRL Press, Oxford.

Thomas, T.C., and M.G. McNamee. 1990. Purification of Membrane Proteins. Meth. Enzymol. 182: 499-520.

West, I., and O. Goldring. 1992. Lectin Affinity Chromatography. pp. 81-89 in Practical Protein Chromatography, A. Kenney & S. Fowell, eds. Meth. Mol. Biol. Vol. 11.

Wisdom, G.B. 1988. Antibody-Enzyme Conjugate Formation. pp.373-382 in New Protein Techniques, J.M. Walker, ed. Humana Press, Clifton, New Jersey.

Yip, T.-T., and T.W. Hutchens. 1992. Immobilized Metal Ion Affinity Chromatography. pp. 17-31 in Practical Protein Chromatography, A. Kenney & S. Fowell, eds. Meth. Mol. Biol. Vol. 11.

11

Chapter 12

Hanging Drop Crystallization

I. Introduction

II. Performing a Hanging Drop Crystallization
    A. Crystallization Principles
    B. Procedure

III. Designing a Follow-up Strategy
    A. Fine-tuning the Sparse Matrix Conditions
    B. Expanding the Initial Screen
    C. Still No Crystals?

IV. Suppliers

V. References

12

I. Introduction

X-ray crystallography is a powerful technique for the study of proteins and macromolecular complexes at atomic resolution. It provides the structural detail required to unravel such aspects of protein function as enzyme mechanisms and ligand binding chemistry. The major drawback of crystallography is the necessity of growing protein crystals that are suitable for X-ray data collection, and this has been a rate-limiting step in many protein structure determinations. Nevertheless, crystallization techniques are becoming more standardized, and it is now recognized that the person who purifies a protein is often the one with the best chance to crystallize it, because he or she is most familiar with its behavior and idiosyncrasies.

Protein crystallization entails systematically varying several solution parameters affecting protein solubility to find the balance between the formation of amorphous precipitates and the growth of crystals. In this chapter, we present the **hanging drop** method for screening large numbers of potential crystallization conditions. This method combines **vapor diffusion** with the use of precipitating agents to bring a protein solution to its **saturation** point in a controlled manner. A small (5 - 20 μl) sample droplet of protein containing buffer and a precipitating agent is suspended from a glass microscope coverslip in a well containing a much larger volume of a second solution of higher osmolarity (Fig. 12.1). Over time, there is a net evaporation of water from the sample droplet that is accompanied by a net condensation into the reservoir solution so as to equalize the osmolarities of the two solutions. This migration of water from the sample droplet results in a concentration of both the protein and the precipitating agent, lowering the solubility of the protein and, if one is lucky, inducing the formation of protein crystals.

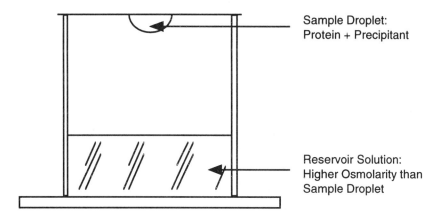

Sample Droplet:
Protein + Precipitant

Reservoir Solution:
Higher Osmolarity than
Sample Droplet

Figure 12.1. The hanging drop method.

Although the hanging drop method is not technically difficult to execute, success at growing crystals suitable for X-ray analysis involves the optimization of many chemical and physical parameters. Consequently, it may seem to the beginner that a dauntingly large amount of theoretical information must be assimilated just to get started. This is not so, and there is much to be gained by "learning along the way." In order that the reader may begin crystallization experiments as quickly as possible, the protocol described in Section II uses commercially available screening kits. Such kits have two advantages: they give the beginner (as well as the experienced crystallizer!) a set of tangible starting points, and they save much of the expense and tedium in preparing large numbers of crystallization solutions.

By focusing on the hanging drop method, we do not wish to discount the usefulness of other methods which exist for crystallizing proteins (described in McPherson, 1982). However, more biological macromolecules have been crystallized by vapor diffusion than by all other methods combined (McPherson et al., 1995), and since the hanging drop method is the most common type of vapor diffusion technique, we feel that it is best starting point for the novice.

12

II. Performing a Hanging Drop Crystallization

A. Crystallization Principles

How do protein crystals grow?

The solubility of a protein in water depends on its amino acid composition and such environmental factors as temperature, pH, and the presence of other solution components. When the concentration of a protein is brought above its solubility limit, the solution becomes **supersaturated.** At this point, the protein begins to **aggregate,** moving from the solution phase to a separate, insoluble phase (Arakawa and Timasheff, 1985). Aggregation occurs in two distinct stages, **nucleation** and **growth.** During nucleation, protein molecules associate into a stable complex either as a disordered, amorphous precipitate or as a microcrystal. Additional protein molecules are then transported to the nucleation complex by diffusion and adhere to its surface in the growth phase.

What conditions favor crystal growth?

The factors influencing the formation of crystals, as opposed to amorphous precipitates, are incompletely understood. Consequently, most crystallizers rely on a systematic variation of factors which affect the solubility of their protein, selecting for those conditions which are found empirically to give rise to the formation of crystals. Amorphous precipitates tend to predominate when the protein concentration is well above saturation. In addition, crystals grow much more slowly than do amorphous precipitates, so if the concentration of a protein is brought above its saturation point too quickly, precipitation again will predominate (Fig. 12.2). Thus, the most general crystallization strategy is to bring the protein as slowly as possible to a point only slightly above its saturation point.

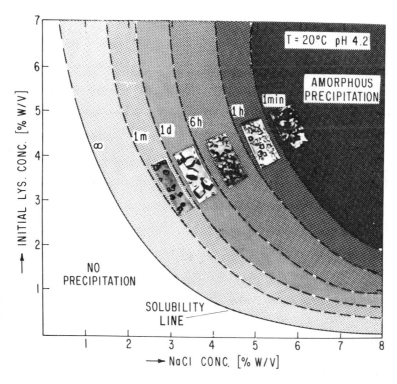

Figure 12.2. Crystallization kinetics of lysozyme as a function of protein concentration and salt concentration. Amorphous precipitates grow fast at high concentrations of salt and protein. Crystals grow successively larger over longer periods at lower concentrations of salt and/or protein. However, below saturation, no crystals or precipitates are observed. Reprinted with permission from Feher and Kam (1985).

12

How is a protein solution brought to its saturation point?

As shown in Figure 12.2, saturation can be achieved either by raising the concentration of the protein, usually by evaporative methods, or by lowering its solubility. Although a number of physical and chemical means are available to lower the solubility of a protein, including varying pH and temperature, the majority of protein crystallizations today employ **precipitating agents**, also called **precipitants**. The hanging drop method is particularly efficient in lowering the solubility of a protein, because it combines evaporation (vapor diffusion) with the use of precipitating agents.

What kinds of precipitating agents are used in crystallization?

Precipitating agents are of two main types, salts and organic molecules, and they work in distinct ways. Salts disrupt the hydration shell of proteins, minimizing attractive protein-solvent interactions and maximizing attractive protein-protein interactions. Organic precipitants function primarily by lowering the dielectric constant of the solution to reduce electrostatic shielding of charged and polar functional groups on proteins (McPherson, 1985a). The net result with either type of precipitant is that attractive interactions between protein molecules are enhanced. This promotes specific bonding which can then lead to crystallization.

Several organic precipitants in particular are useful because they do not denature proteins as many other organic precipitants do. These precipitants are 2-methyl-2,4-pentanediol (MPD) and the poly(ethylene glycol)s (PEGs). Poly(ethylene glycol)s form a special category of organic precipitants, because they are very large polymers. While their function is partly linked to their ability to lower the dielectric constant of the solution, they also appear to affect the structure of water (McPherson, 1985b). PEGs are produced as a mixture of ethylene glycol chains of varying lengths which are then fractionated into molecular weight intervals of several hundred or several thousand and sold according to the average molecular weight of the fraction (with average M_r from 200 to 20,000). (Note: PEGs may be contaminated with aldehydes and peroxides as well as a number of anionic species. Several procedures exist for purification of PEGs (Ray and Puvathingal, 1985; Lorber and Giegé, 1991), and crystallization grade PEGs are commercially available in several sizes (Hampton Research).) Recently, a number of other "polymeric precipitants" have been found to be useful for protein crystallization, and this is a class of precipitants that will probably continue to grow (Patel et al., 1995).

A list of some precipitating agents used in protein crystallization is given in Table 12.1. The highest numbers of macromolecular crystals have been obtained using, in descending order of frequency, ammonium sulfate, poly(ethylene glycol)s, Na/K phosphate, sodium chloride, MPD, and magnesium chloride (McPherson et al., 1995). It should be noted, though, that MPD has come into use relatively recently, so it may be underrepresented in this ranking. Naturally, it is possible to combine several precipitants in a single crystallization experiment.

Table 12.1
Precipitating Agents Used in Protein Crystallization
From McPherson (1985a), Gilliland and Davies (1983), Patel et al. (1995)

Salts	Organic precipitants
Ammonium acetate	Acetone
Ammonium citrate	Dioxane
Ammonium nitrate	Dimethyl sulfoxide
Ammonium sulfate	*tert*-Butanol
Citrate	Ethanol
Cadmium sulfate	Glucose
Lithium chloride	Glutamic acid
Lithium sulfate	Isopropanol
Magnesium chloride	Methanol
Magnesium sulfate	2-Methyl-2,4-pentanediol
Phosphate	1,3-propanediol
Potassium borate	*n*-Propanol
Potassium nitrate	
Potassium phosphate	**Polymeric precipitants**
Potassium sodium phosphate	Poly(acrylic acid) (various M_r)
Potassium tartrate	Carboxymethylcellulose
Sodium chloride	Poly(ethylene glycol) (various M_r)
Sodium citrate	Poly(ethylene glycol) 2000
Sodium iodide	dimethyl ether
Sodium nitrate	Poly(propylene glycol) P400
Sodium phosphate	Poly(vinyl alcohol) 15,000
Sodium sulfate	Poly(vinylpyrrolidone) K15

12

How do I find crystallization conditions for my protein?

Protein crystallization experiments generally proceed in two stages. First, a broad range of conditions is screened to test the solubility of the protein with regard to precipitating agents and other solution components. However, it is rare to obtain diffraction-quality crystals at this stage. Instead, it is hoped that boundaries can be established between conditions in which the protein is soluble and those in which it is insoluble . Usually at this stage, insoluble protein is observed as an amorphous precipitate. This means that the precipitation conditions are too severe to allow crystal growth (see Fig. 12.2). Thus, in the second, "optimization" stage of the experiment, conditions which gave rise to precipitates in the first stage are modified systematically in small increments to allow the more gradual approach to insolubility that is required for the formation of crystal nuclei (Cudney et al., 1994).

The protocol presented in this section is an example of a first-stage screen, while the development of optimization strategies is discussed in Section III.

12

B. Procedure

The following protocol makes use of the **sparse matrix** approach to screening a large number of possible crystallization conditions on the basis of pH, precipitating agent, and type and concentration of buffer and salt components (Jancarik and Kim, 1991). This method defines a set of precipitant-buffer solutions as the starting point for crystallization screens. It is based on the assumption that the best conditions for screening are those which have been used successfully for other proteins, especially if they are structurally related to the one of interest (Cudney et al., 1994). Two separate kits containing appropriate sparse matrix screening solutions are available from Hampton Research (Section V).

Hanging drop crystallizations are normally carried out in 24-well tissue culture plates (Fig. 12.3). These plates allow for easy organization, storage, and examination of crystallization experiments. Sample droplets are suspended from glass microscope coverslips, which are sealed to the wells with vacuum grease. Several preparatory steps (Sections II.B.1 and II.B.2) need to be carried out prior to setting up the experiment itself.

1. Preparing Coverslips

Protein adheres to and wets glass surfaces. This wetting leads to flattening of the sample droplet during a crystallization experiment, limiting the ultimate size to which crystals can grow. Flattening also can lead to uneven evaporation and excessive exposure of the protein to air because of the increased surface area of the droplet. A second problem is that the protein crystals themselves can adhere to the coverslip, making them susceptible to damage when attempts are made to dislodge them. To alleviate these problems, glass coverslips are usually coated with a film of organosilane prior to use in hanging drop experiments.

This process can be carried out at any point prior to the crystallization.

12

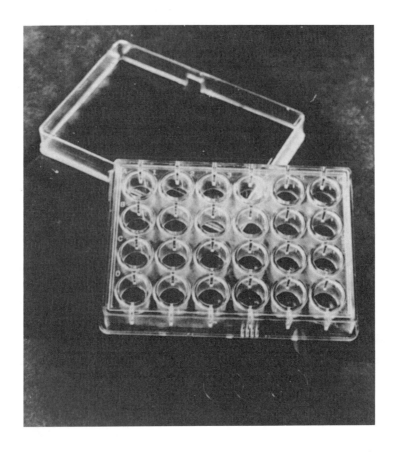

Figure 12.3. Tissue culture plate used for hanging drop experiments. Reprinted with permission from McPherson (1985b).

12

- **Equipment**
  1. 2.2 cm diameter round glass coverslips
  2. Forceps
  3. Whatman 3MM filter paper

- **Reagents**
  1. PROSIL-28 organosilane formula
  2. Sodium hydroxide
  3. MICRO glass cleanser

- **Protocol**
  1. Wash coverslips by soaking overnight in dilute sodium hydroxide or dilute MICRO cleanser. Rinse thoroughly with water and allow to dry.

  2. Dilute 1 ml PROSIL-28 into 100 ml water.

  3. Immerse clean, dry coverslips into diluted PROSIL-28 solution for one minute. Coverslips can be stood on edge on the bottom of the beaker during this time, or they can be placed in a coverslip holder.

  4. Remove coverslip from PROSIL-28 solution and rinse under deionized, distilled water for one minute.

  5. Lay coverslips on a large sheet of Whatman 3MM filter paper. Blot the upper surfaces dry with a second sheet of filter paper. Allow to dry at room temperature, then cure overnight in a drying oven at 100°C.

- **Comments**
  1. Silane-coated ("siliconized") coverslips are commercially available.

  2. Silane-coated coverslips should be stored in a dust-free container. They are stable indefinitely at room temperature.

  3. Silane-coated coverslips can be autoclaved to ensure sterility.

  4. Diluted PROSIL-28 solution is stable for 24 hours. Discard when a white precipitate appears.

  5. PROSIL-28 concentrate is sensitive to moisture. Once opened, it should not be kept more than about a month unless it is stored desiccated, under which circumstances it may be stable for a year or longer. In any case, do not use when a precipitate is present.

12

2. Preparing Protein

It is assumed that the protein to be crystallized has been purified as much as possible. Although there are no absolute criteria, a general rule of protein crystallization is: the purer the protein, the better your chances will be of crystallizing it. Purity should be assessed by two-dimensional electrophoresis, employing either isoelectric focusing or non-equilibrium pH-gradient electrophoresis in the first dimension (Chapter 7). One-dimensional SDS-PAGE (Chapter 5) may not be sufficient if contaminants include protein isotypes or post-translationally modified forms which have molecular weights similar to the protein of interest. Ideally, protein should be crystallized immediately after purification to minimize functional and structural degradation due to oxidation, proteolysis, etc., but this is not always possible. If you need to store your protein prior to crystallization, use the guidelines in Chapter 1, Section VI.C.

• **Protocol for preparing protein for crystallization**

This procedure should be carried out immediately before the crystallization experiment is to begin. In general, the protein is transferred to a starting buffer, concentrated, and centrifuged to sterilize and to remove any pre-existing protein aggregates. The starting buffer should be at a concentration of 5 mM or less so that it does not become a significant component of the sample droplet.

1. All solutions containing protein, buffer, or precipitating agents should be made using the highest purity water that is available. *The importance of water quality cannot be over-emphasized*. Most crystallography laboratories invest in their own private water purification systems such as the Millipore Milli-Q system. For best results, do not store water in large carboys, since they become contaminated over time. Instead, take water directly from the delivery system just prior to use.

2. Transfer protein to a suitable buffer (see Table 12.2 for examples) by dialysis (Chapter 4) or gel filtration (Chapter 10) using Bio-Gel P-6DG or Sephadex G-25. For small volumes (1 - 2.5 ml), ready-made columns such as Pharmacia's PD-10 are convenient. If starting with ammonium sulfate-precipitated material, dialyze or dilute as needed to solubilize.

3. Centrifuge protein 10 min at 20,000 x g and 4°C to remove denatured protein which could clog ultrafiltration membranes in subsequent concentration steps.

4. Concentrate protein to 5 - 30 mg/ml if protein is soluble at that level (Chapter 4).

5. Recentrifuge to remove any aggregated material that may form during concentration. For additional security against bacterial contamination, filter with a 0.22 μm pore-size membrane and transfer protein to a sterile container. If protein is lost after filtration, try a pore size of 0.45 or 0.65 μm; however significant protein loss may signify instability and preliminary aggregation.

•   **Comments**
    1. The best protein concentration for crystallization is usually 5 - 30 mg/ml (McPherson, 1982).

    2. Biologically inert buffers, such as Good's buffers, are usually the best for crystallization (Table 12.2).

12

Table 12.2
Common Buffers for Crystallization

Buffer	pK$_a$	pH Range
Acetate	4.76	4.2 - 5.3
Citrate	5.40 (pK$_{a3}$)	4.9 - 5.9
MES	6.15	5.6 - 6.7
Imidazole	6.95	6.4 - 7.5
HEPES	7.55	7.0 - 8.1
Tris	8.06	7.5 - 8.6
Bicine	8.35	7.8 - 8.9

3. Setting Up the Crystallization

- **Equipment**
  1. Tissue culture plates, 24-well, sterile
  2. Modeling clay (plasticene)
  3. Silane-coated ("siliconized") 2.2 cm-diameter round coverslips (Section II.B.1)
  4. Vacuum grease
  5. 10 ml syringe with #10 gauge needle
  6. Flat-nosed forceps
  7. Stereomicroscope

- **Reagents**
  1. Prepared protein solution (Section II.B.2)
  2. Hampton Crystal Screen Sparse Matrix Kit
  3. Hampton Crystal Screen II Sparse Matrix Kit (optional)

12

- **Protocol**
  1. Unwrap a 24-well tissue culture plate. Mark the bottom of each well with a number (1 through 50; or more if using both screening kits) corresponding to a solution in the Crystal Screen kit.

  2. Remove the lid. Make four 5 mm-diameter balls out of the modeling clay. Press one ball into each of the four corners of the inside surface of the base as shown in Figure 12.4. These balls will act as bumpers to protect the wells from the weight of the lid as well as to keep the coverslips from sticking to the lid and being pulled off whenever the lid is removed from bottom. (If clay is not available, use any material that will accomplish the same effects.)

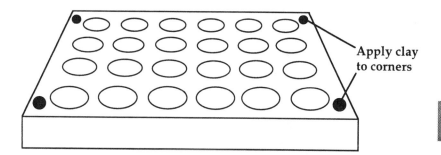

Figure 12.4 Lid bumpers

  3. To each well, add 1.0 ml of appropriate precipitant-buffer solution.

  4. Fill the 10 ml syringe with vacuum grease.

5. Using the #10 gauge needle, trace a line of grease about 2 mm thick around the entire top rim of each well. The circle should be continuous; do not allow any breaks in it. If you find it too difficult to force the grease through the needle, you can remove it from the syringe. It will require somewhat more care to control the grease from spilling into the well and contaminating the well solution.

6. Lay a coverslip on a clean towel or a piece of filter paper, making sure to handle it only at the edges.

7. Pipette 5 µl of protein solution onto the center of each coverslip (see comment 1).

8. Add appropriate precipitant-buffer solution as follows: Have two pipettors ready, one set at 5 µl and the other set at 9 µl. Pipette 5 µl of the correct precipitant-buffer solution onto the protein droplet to create one large drop. Immediately use the second pipettor (the one set at 9 µl) to draw the entire droplet up and down 2 or 3 times. This ensures thorough mixing of protein and precipitant; for viscous protein and/or precipitant solutions, 5 to 10 times may be needed. The second pipettor is set at 9 µl instead of 10 to avoid uptake of air into the pipettor. Be careful to avoid introducing bubbles during this process. If any bubbles remain in the sample droplet, remove them with the pipettor.

9. Pick up the coverslip by the edge, using the forceps. Turn it over so that the droplet hangs upside down. Place the coverslip on the appropriate well so that its outer edge is centered over the rim of the well.

10. Press the coverslip down onto the top of the well by pushing with fingertips simultaneously at four corners around the edges. Press firmly enough to squeeze the coverslip as close to the well as possible, but not so hard so as to break it.

12

11. Continue in the same way with the remaining wells. When all wells are completed, gently place the lid on the plate. The bumpers should prevent the lid from resting on the coverslips.

12. Once crystallization plates are set up, they can be stacked conveniently on top of one another for storage at the desired temperature (see comment 7, and discussion in Section III.A.2). Choose a quiet, out-of-the-way area with minimal temperature fluctuations. Storage on stationary shelves is preferable to sliding drawers, since the plates are less likely to be subjected to jarring.

• **Comments**
1. Sample droplets of between 5 and 20 µl (e.g., between 2.5 and 10 µl protein) are acceptable. In early screening stages, it is best to use smaller volumes to conserve protein. In later stages, larger volumes are more conducive to growing large crystals.

2. Check each plate under a stereomicroscope after it is completed and before storing the plate. Often you will get a good indication of which conditions render the protein insoluble even at this early stage.

3. While turning the coverslips over, ensure that the sample droplet remains centered on the coverslip.

12

4. After the coverslip has been placed on a well, there should be an even barrier of vacuum grease visible under the coverslip. If there are breaks or gaps in the barrier, rotate the coverslip slightly to spread the vacuum grease. Any gaps that are left will allow vapor to leak out, and may lead to total evaporation and drying of the well.

5. Ensure that the protein and precipitating agent are mixed completely to ensure a homogeneous, monodisperse solution.

6. Bacterial contamination should not be a problem when using high salt concentrations as precipitating agents, but it can be a problem with organic precipitants such as PEG. Be sure that the protein solutions are sterilized by filtration. Precipitant solutions obtained commercially are sterile; take care to use aseptic techniques when opening them. Tissue culture plates are also sterile, but you may want to autoclave pipette tips and coverslips. Small amounts of microbial inhibitors such as 0.02% NaN_3 (Chapter 1, Section II.E.) can be included in solutions, but be aware that they could affect the crystallization if they interact with the protein. Also, azide must never be used in crystallization solutions that contain heavy metal ions because of the danger of forming explosive metal azide salts (Rozycki and Bartha, 1981).

7. If resources permit, set up a duplicate of each plate, so that one can be stored at room temperature and the other can be stored in a refrigerator or cold room (4°C).

4. Examination of Plates

Since protein crystals can be unstable and have been known to disappear as suddenly as they appear, it is important to record the appearance of each well on a regular basis. Observe them immediately after setting up your crystallization experiment, and again the following day. Make another set of observations after two to three days, and thereafter once a week. Hanging drop experiments are best examined with a stereomicroscope. Be careful when moving plates for examination--excessive shaking or jolting can limit the overall size of the crystals.

What should I look for?

Protein used for crystallization experiments should be devoid of denatured, aggregated protein (Section II.B.2) and so should be transparent with a slight bluish tint. After mixing with precipitant-buffer solution, protein droplets in which amorphous precipitates develop will have the same appearance as any other solution containing insoluble protein. Initially, there may be a uniform, light gray or white turbidity. This may appear within minutes or hours of mixing together protein and precipitate, or it may not be apparent for several days. In either case, over time the precipitate will usually clump to form large masses. Some precipitants will give a precipitate with a more unique appearance, so the same protein may not look the same in every well of your sparse matrix screen.

Crystals are usually easy to distinguish from clumps of amorphous precipitate. Crystals are transparent and have definite form characterized by planar faces (Fig. 12.5), while precipitate clumps are opaque and have irregular shapes without defined edges and faces (Fig. 12.2). In addition, crystals are often **birefringent**, so that they appear alternatingly bright and dark as they are rotated under crossed polarizers in the microscope.

In some cases, you will find entities which look too organized to be simple precipitates, but which are clearly not truly crystalline. These are sometimes due to growth of crystals among amorphous precipitates, so that the crystals are covered with precipitate. On the other hand, they may be **precrystalline aggregates**, which are often an indication that you are close to, but not quite at, the right solution conditions for crystallization. In this case, it may take only a small variation of your screening conditions to obtain crystals.

Is everything I see interesting?

No. At least, not in a crystallographic sense. At some point, probably every crystallographer has had the experience of finding a glass or textile particle which appears to have

12

straight edges and is highly birefringent--a gold mine for a forensic microscopist, perhaps, but not much use to a protein crystallographer! Experience and reproducibility are guides in identifying such contaminants.

I have crystals, but are they protein?

Buffers in hanging drop experiments are generally used at low concentrations and are not likely to crystallize. However, some salt precipitants, such as ammonium sulfate and the potassium sodium phosphates are used at high enough concentrations that they may crystallize instead of the protein. This is especially true if salts are mixed with precipitating ions (*e.g.*, phosphate buffer with Ca^{2+}) or with organic molecules which may change the solubilities of all solutes.

12

Figure 12.5. Sampling of the many forms that protein crystals can take. (A) deer liver catalase, (B) turkey liver fructose-1,6-diphosphatase, (C) guinea pig cortisol-binding protein, (D) concanavalin B from Jack Bean, (E) beef liver catalase, (F) protein of unknown function from pineapple, (G) *E. coli* elongation factor Tu, (H) yeast phenylalanine tRNA, (I) Gene 5 DNA unwinding protein from bacteriophage fd, (J) chicken muscle glycerol-3-phosphate dehydrogenase, and (K) canavalin from Jack Bean. Reprinted with permission from McPherson (1982).

12

In most cases, such salt crystals appear quickly and grow rapidly, while protein crystals usually grow over a period of weeks or months or even longer. Several methods are available to test whether your crystals are protein or salt. They are listed in order of increasing reliability:

- *Crush test.* Carefully disassemble the well by removing the coverslip and placing it droplet side up on the stereo-microscope. Using a fine glass needle (made from a 100 μl glass capillary that has been heated in the middle and then drawn apart) or a micromanipulation tool, crush a sample crystal. Salt crystals usually crush with difficulty into relatively few pieces, while protein crystals usually crush easily into a shower of very fine pieces.

  Comment 1: Wells can be reassembled after removing a crystal to perform the crush test. Be careful not to leave the droplet exposed to air any longer than is necessary.

  Comment 2: Protein crystals generally contain a high proportion of solvent. It is very important to keep them wet in either the original sample droplet or a medium, such as the lower well solution, that maintains the correct osmolarity to keep them from dissolving. Bear in mind, though, that the washing medium should also contain any cofactors or ligands that are present in the sample droplet, but may not have been added to the lower well solution. This comment applies to all manipulations of crystals after they are removed from the original crystallization well.

- *Dehydration Test:* Allow a single crystal to dehydrate in air. Protein crystals will usually disintegrate, while salt crystals usually dry intact (McPherson, 1982).

- *Gel electrophoresis.* If your crystals are large enough, you can solubilize a single crystal and analyze it by SDS-PAGE (Chapter 5). A cubic crystal that is 0.5 mm on each side has a volume of 0.125 µl and probably contains about 60 µg of protein. This quantity of protein is easily detectable by SDS-PAGE followed by silver staining (Chapter 5). If your crystals are very small (0.05 µl or smaller), you will need to use a large number of them. In either case, place the crystals in a 1.5 ml centrifuge tube with about 100 µl of the solution in the lower well (but see comment 2 for the crush test above), and centrifuge at 10,000 x g for 15 min. Wash the crystal(s) with another 100 µl of lower-well solution and recentrifuge. Suspend the crystal(s) in 20 µl SDS-PAGE sample buffer and carry out denaturing gel electrophoresis and silver-staining as described in Chapter 5.

  Comment 1: It is important to wash the crystals thoroughly to remove soluble protein in the crystal-bathing solution. The larger the crystal, the higher the "signal-to-noise" ratio, and the more likely it is that the appearance of protein on the gel originates from the crystal itself.
  Comment 2: Bear in mind that crystals may dissolve while washing them prior to SDS-PAGE.

- *Dye binding*: Hampton Research markets the "Izit Crystal Dye." A small amount added to the sample droplet will give a blue color to protein crystals but not to salt crystals.

- *X-ray diffraction.* This is the ultimate test, since crystalline periodicities characteristic of proteins are readily distinguishable from those of salt crystals. A description of X-ray crystallography is beyond the scope of this text, and the reader is referred to the references in Section V.

12

III. Designing an Optimization Strategy

This section discusses how to optimize your crystallization strategy around the results obtained from sparse matrix screening of precipitation conditions. At this point, you have completed your initial screen as described in Section II and have obtained one of three results:

1. You obtained crystals of your protein with one or more of the conditions in the sparse matrix kits.

2. You observed amorphous precipitates or precrystalline aggregates of some type with one or more of the conditions in the sparse matrix kits.

3. You observed no crystals, precipitates, or aggregates with any of the conditions in the sparse matrix kits.

If you obtained results 1 or 2, you may want to fine-tune your crystallization experiment using the guidelines in Section A below. If you obtained result 3, proceed to Section B.

A. Fine-tuning the Sparse Matrix Conditions

1. A Practical Example

The sparse matrix screen has yielded a number of precipitant-buffer conditions in which your protein is insoluble. What is now required is to design a narrow-range **grid screen** for each of those conditions by varying the pH and the concentrations of each component systematically and observing whether one or more of those variations gives good crystals. In addition, this would be the time to begin to include a test of the effects of ligands on the crystallization of your protein.

Because you may have a large number of conditions to test, it is best to start with those that already have yielded crystals and then move to those containing precrystalline aggregates and amorphous precipitates. It is worthwhile to test as many conditions as is practical, even if crystals are observed early on, because often multiple crystal forms or crystallization conditions can be helpful at later stages of crystallographic analysis.

As an example of the screening process, suppose you found that condition # 35 of the Hampton Crystal Screen (0.1 M HEPES, pH 7.5; 1.6 M Na/K phosphate) gave an amorphous white precipitate after 1 day at room temperature. You might then set up a grid screen in a single 24 well tissue culture plate using 0.8, 1.0, 1.2 and 1.4 M Na/K phosphate with a pH gradient of 7.0, 7.2, 7.4, 7.6, 7.8 and 8.0 (Fig. 12.6). Notice that the pH is chosen to bracket that of the Crystal Screen combination, but the concentrations of precipitating agent (Na/K phosphate) are graduated below that used in the initial screen.

It may take several cycles of fine-tuning refinement to find the right balance of conditions favoring crystallization over formation of amorphous precipitate. *Patience is important*: crystal growth is usually much slower than precipitation, and while precipitation may be apparent after only minutes of incubation, crystal growth can take anywhere from hours to weeks (Fig. 12.2). If you know any crystallographers personally, you are bound to hear at least one story of an apparently unsuccessful crystallization experiment that was forgotten, only to turn up years later with beautiful crystals in it!

12

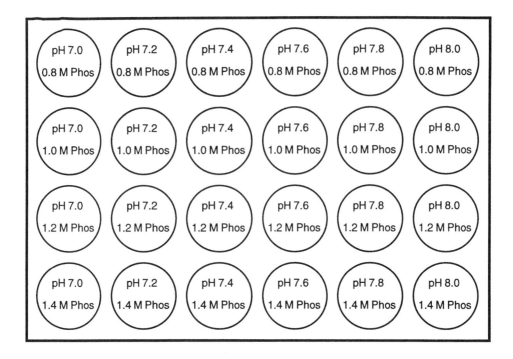

Figure 12.6. Grid screen to fine-tune Crystal Screen®
condition # 35

12

2. A General Guide

As discussed above, protein crystallization entails systematically varying parameters to find the balance between amorphous precipitation and crystallization. The following factors are particularly important in optimizing your crystallization strategy:

pH and Buffer

The formation of protein:protein contacts is the basis for the assembly of a protein into a crystal lattice. These contacts are highly dependent on local charge distribution, which is physically complex and responds to the dielectric constant of the solution, pH, and conformational variations within the protein (Honig and Nicholls, 1995). Thus, pH is almost always a critical parameter in finding crystallization conditions for a protein.

Many salts and organic solvents commonly used as precipitating agents (Table 12.1) do not have significant buffering capacity, so buffers are usually included as additional components. Buffers used during protein purification, such as Good's buffers (Good et al., 1966), are usually also suitable for crystallization. However, since it is important to use a buffer whose pK is close enough to the desired pH to provide significant buffering capacity (Chapter 1), it is necessary to switch between buffers when screening a wide pH range (Table 2.2). A reasonable concentration range for buffers is between 50 and 200 mM; note that this buffer concentration range is much higher than that used for the preparation of the protein described in Section II.B.2.

The majority of protein crystals reported in the literature grow at a pH of 7.0, with a bell-shaped decrease in frequency at higher and lower pH values (McPherson et al., 1995). Nevertheless, enough crystals are observed at extreme pH values (below 5.0 and above 9.0) that it is worth including a search over a wide range of pH values in your optimization strategy. This is true even if you are starting from one of the conditions from the sparse matrix screen, although you will probably want to start with a narrower range to bracket the pH at which crystals or precipitate is first observed. Many crystallizers start with a very broad pH range, setting up experiments at pH 4.0, 5.0, 6.0, 7.0, 8.0 and 9.0 for a given combination of protein and precipitant. In particular, you may want to test pH values near the functional isoelectric point of your protein to take advantage of the fact that most proteins show a decreased solubilities near isoelectric pH.

Although it is difficult to predict which pH will be the best for crystallizing a given protein, *reproducibility* of pH measurements is extremely important for a well-controlled comparison of crystallization parameters between experiments. You may find that even a change of a tenth of a pH unit can significantly affect the success of a crystallization. With this in mind, it may be helpful to follow these guidelines to safeguard the reproducibility of your pH measurements:

- If possible, dedicate a pH electrode to be used exclusively for crystallization work. This eliminates variations between electrodes and minimizes the introduction of unknown contaminants into your solutions.

- Do not immerse your pH electrode into your entire crystallization solution. Instead, measure pH from small aliquots, and then discard them. Most combination pH electrodes release reference ions into the solution being measured (usually K^+ and Cl^-, but sometimes also Ag^{2+}), thereby contaminating them.

- Use the pH meter to *confirm,* rather than to *adjust* the pH of a solution. Add known amounts of reagents such as KOH or HCl instead of titrating with a dropper, so that in future preparations the pH can be adjusted by adding exact proportions of reagents instead of relying on the reproducibility of the electrode.

Temperature

More macromolecules have been crystallized in the temperature range of 0-4°C (almost 1400) than in any other temperature range (McPherson et al., 1995). Crystallization at low temperature has several advantages. Most proteins are less soluble at 4°C than at room temperature (McPherson, 1982). In addition, many proteins are more stable at 4°C than at room temperature, so low temperatures help to protect the protein during the extended waiting periods typical of crystallization experiments. A final consideration is the rate of solubility change of the protein. In hanging drop experiments, both protein and precipitant are concentrated three to five times faster at 20°C than at 3° (Mikol et al., 1990). The slower change in protein solubility at low temperature gives it more time to aggregate into ordered crystalline nuclei.

Despite these considerations, it may simply be more convenient to work at room temperature if cold-room or refrigerator storage space is in short supply, or to avoid the complications of temperature-induced changes in solution pH. Almost 800 macromolecules have been crystallized in the temperature range of 20 - 24°C (McPherson et al., 1995). The stability of your protein during crystallization will be an important factor.

Concentrations of Protein and Precipitating Agent

The driving force in a hanging drop crystallization is the difference in osmolarity between a droplet containing the protein and precipitating agent and a higher osmolarity reservoir whose volume is several orders of magnitude larger than that of the protein sample (Fig. 12.1). During the experiment, the concentrations of protein and precipitant slowly increase as water vapor leaves the sample droplet and condenses into the reservoir. Initial conditions are set so that the volume decrease in the sample droplet is about 25 to 50% when the system reaches equilibrium, usually after between one day and one week (Mikol et al., 1990). During this time the protein solution starts out below its saturation point, and then slowly pass through and above it.

If no precipitant were present in the protein solution, the best approach would be to determine the solubility of the protein by concentrating it until it precipitated, as evidenced by visible turbidity or measured light scattering. The protein could then be diluted to a concentration perhaps 25% below its saturation point and sealed in the vapor diffusion chamber and would slowly re-concentrate as the system reached equilibrium. However, the inclusion of a precipitating agent with the protein makes the system more complex. There is now a range of saturation points that depends on the concentration of precipitating agent, and it is necessary to balance the concentrations of protein and precipitating agents to be used in a given experiment.

In practice, the concentration of protein is usually initially chosen to be as high as it is possible to maintain in the absence of precipitating agents. Generally, the best concentration is considered to be between 5 and 30 mg/ml, although proteins have been crystallized at concentrations as low as 1 mg/ml and as high as several hundred (McPherson, 1982). The concentrations of precipitating agents are then varied until the protein begins to precipitate.

It is also often useful to vary the concentration of protein along with the concentration of precipitating agent. This may be helpful when even low concentrations of precipitant lead to precipitation of protein or when crystal defects such as twinning are a problem. Keep in mind though, that lowering the protein concentration below the optimal range may lead to growth of smaller crystals.

Whether it is fixed or varied along with other parameters, protein concentration should be as reproducible as any other component of your crystallization solution. Variations in protein concentration will affect the rate of approach to saturation (Fig. 12.2). Be especially careful to account for substances which interfere with protein concentration determinations (Chapter 3), especially those that are likely to vary from experiment to experiment.

Ligands and Other Solution Components.

Metal ions, nucleotides, peptides, enzyme substrates, and cofactors are all potentially important factors to be considered for inclusion in crystallization experiments. Not only are ligands and cofactors often useful in stabilizing proteins, they can even have significant structural effects which affect the crystallization of proteins. An extreme example of this is actin, which cannot be crystallized as a pure protein because of its tendency to polymerize into helical filaments. Actin can only be crystallized in the presence of protein ligands such as deoxyribonuclease I, profilin, or gelsolin (Mannherz, 1992) which interfere with polymerization and keep actin in its monomeric form. Another example is elongation factor Tu, which exhibits different crystal forms depending on whether it is bound to GDP or an analogue of GTP (Berchtold et al., 1993).

If you are not sure about the ligand requirements of your protein and you have information about its amino acid sequence, it may be helpful to test whether it is structurally related to other proteins with known ligand-binding capacity. For example, a recent homology search identified six proteins as being structurally homologous to flavodoxin and predicted that they contained binding sites for flavine mononucleotide (Grandori and Carey, 1994).

12

In addition to specific ligands and cofactors, other solution components such as detergents and protein stabilizing agents have been found to improve the crystallizability of many proteins (McPherson et al., 1986; Sousa, 1995).

B. Expanding the Initial Screen

It is possible that none of the combinations from the Crystal Screen or Crystal Screen II kits gave evidence of precipitation or aggregation with your protein. Or, perhaps your attempts to fine-tune the results of your initial screen failed to identify conditions which give rise to good quality crystals. In either case, you need to expand your initial screen from the first 50 or 100 conditions defined by the Crystal Screen kits. This can be done by either setting up a number of grid screens (Section III.A) to sample a small number of precipitants or by using other sparse-matrix screens described in the literature (cf. Cudney et al., 1994). Unfortunately, alternative sparse matrix screens are not yet commercially available, although a number of grid screening kits are available from Hampton Research.

C. Still No Crystals?

The screening strategies presented in this section are meant to convey the rudiments of protein crystallization, but they cannot guarantee success. An in-depth discussion of optimization strategies for crystallization is beyond the scope of this book, and the reader is referred to the texts in Section V for further details.

What we can say here is that, more than any other technique described in this book, success at crystallization is as often a result of the creativity of the investigator as from the following of established methods. You may have an intuition that a certain cofactor, chemical modification, or mutation will stabilize your protein or affect its solubility sufficiently to tease it towards crystallization. For example, the long-awaited structure of the S1 head of myosin was possible only after exhaustive, reductive methylation of its lysine residues (Rayment et al., 1993). Fortunately, the vast majority of protein crystallizations do not require such drastic measures.

IV. Suppliers

Coverslips: Fischer, Hampton Research, Thomas, VWR (unsiliconized); Hampton Research (siliconized)

Crystal Screen I® & II® Sparse Matrix Kits: Hampton Research

Chromatography paper, Whatman® 3MM: Fisher, Thomas, VWR

MICRO® cleanser: International Products

Millex® Filters: Millipore

Modeling clay (plasticene): Hampton, any artist's supply store

MPD (2-methyl-4-pentanediol) (crystallization grade): Hampton Research

Poly(ethylene glycol): Aldrich, Fluka, Hampton Research (crystallization grade: available only for M_r 1000, 6000, 8000), Sigma

PROSIL-28® Organosilane Formula: Hampton Research

Tissue Culture Plates: Fisher, Hampton Research, Sigma, Thomas, VWR

Vacuum Grease (Dow Corning®): Fisher, Hampton, Thomas, VWR

12

V. References

Arakawa, T. and S. N. Timasheff. 1985. Meth. Enzymol. 114: 49-77. Theory of Protein Stability.

Berchtold, H., L. Reshetnikova, C. O. A. Reiser, N. K. Schirmer, M. Sprinzl and R. Hilgenfeld. 1993. Nature 365: 126-132. Crystal Structure of Active Elongation Factor Tu Reveals Major Domain Rearrangements.

Cudney, B., S. Patel, K. Weisgraber, Y. Newhouse and A. McPherson. 1994. Acta Cryst. D50: 414-423. Screening and Optimization Strategies for Macromolecular Crystal Growth.

Gilliland, G. L. and D. R. Davies. 1983. Meth. Enzymol. 104: 370-381. Protein Crystallization: The Growth of Large-Scale Single Crystals.

Good, N. E., G. D. Winget, W. Winter, T. N. Connolly, K. M. Izana and R. M. M. Singh. 1966. Biochemistry 5 : 467-477. Hydrogen Ion Buffers for Biological Research.

Grandori, R. and J. Carey. 1994. Protein Sci. 3: 2185-2193, Six New Candidate Members of the α/β Twisted Open-sheet Family Detected by Sequence Similarity to Flavodoxin.

Honig, B. and A. Nicholls. 1995. Science 268: 1144-1149. Classical Electrostatics in Biology and Chemistry.

Jancarik, J. and S. H. Kim. 1991. J. Appl. Cryst. 24: 409-411. Sparse Matrix Sampling: A Screening Method for Crystallization of Proteins.

Lorber, B. and R. Giegé. 1991. Preparation and Handling of Biological Macromolecules for Crystallization. In A. Ducruix and R. Giegé, eds. Crystallization of Nucleic Acids and Proteins: A Practical Approach. 331 pages. IRL Press, Oxford University Press, Oxford.

Mannherz, H. G. 1992. J. Biol. Chem. 267: 11661-11664. Crystallization of Actin in Complex with Actin-Binding Proteins.

McPherson, A. 1982. Preparation and Analysis of Protein Crystals. 371 pages. Wiley Interscience, John Wiley and Sons, New York

McPherson, A. 1985a. Meth. Enzymol. 114: 112-120. Crystallization of Macromolecules: General Principles.

McPherson, A. 1985b. Meth. Enzymol. 114: 120-125. Use of Polyethylene Glycol in the Crystallization of Macromolecules.

McPherson, A., S. Koszelak, H. Axelrod, J. Day, R. Williams, L. Robinson, M. McGrath and D. Cascio. 1986. J. Biol. Chem. 261: 1969-1975. An Experiment Regarding Crystallization of Soluble Proteins in the Presence of β-Octyl Glucoside.

Mikol, V., J.-L. Rodeau and R. Giegé. 1990. Anal. Biochem. 186: 332-339. Experimental Determination of Water Equilibration Rates in the Hanging Drop Method of Protein Crystallization.

Patel, S., B. Cudney and A. McPherson. 1995. Biochem. Biophys. Res. Comm. 207: 819-828. Polymeric Precipitants for the Crystallization of Macromolecules.

Ray, W.J. and J. Puvathingal. 1985. Anal. Biochem. 146: 307-312. A Simple Procedure for Removing Contaminating Aldehydes and Peroxides from Aqueous Solutions of Polyethylene Glycols and of Nonionic Detergents that are Based on the Polyoxyethylene Linkage.

Rayment, I., W. R. Rypniewski, K. Schmidt-Bäse, R. Smith, D. R. Tomchick, M. M. Benning, D. A. Winkelmann, G. Wesenberg, H. M. Holden. 1993. Science 261: 50-58. Three-dimensional Structure of Myosin Subfragment-1: A Molecular Motor.

Rozycki, M. and R. Bartha. 1981. Appl. Env. Microbiol. 41: 833-836. Problems Associated with the Use of Azide as an Inhibitor of Microbial Activity in Soil.

Sousa, R. 1995. Acta Cryst. D51: 271-277. Use of Glycerol, Polyols and other Protein Structure Stabilizing Agents in Protein Crystallization.

12

Crystallization

Carter, C. W., editor. 1990. *METHODS: A Companion to Methods in Enzymology*, Vol. 1. *Protein and Nucleic Acid Crystallization*. Academic Press, Inc., San Diego.

Ducruix, A. and R. Giegé, editors. 1992. *Crystallization of Nucleic Acids and Proteins: A Practical Approach*. 331 pages. Oxford University Press, Oxford.

Hampton Research. *Crystallization Research Tools* (catalogue and technical bulletin). Three issues per year.

McPherson, A. 1982. *Preparation and Analysis of Protein Crystals.* 371 pages. Wiley Interscience, John Wiley and Sons, New York.

McPherson, A., A. J. Malkin and Y. G. Kuznetsov. 1995. *Structure* 3: 759-768. The Science of Macromolecular Crystallization.

McRee, D. E. 1993. *Practical Protein Crystallography.* 386 pages. Academic Press, Inc., San Diego.

Wyckoff, H. W., C. H. W. Hirs and S. N. Timasheff, editors. 1985. *Meth. Enzymol.* Vol. 114: *Diffraction Methods for Biological Macromolecules.* Academic Press, Inc., Orlando.

X-Ray Diffraction

Blundell, T. L. and L. N. Johnson. 1976. *Protein Crystallography.* 565 pages. Academic Press, Inc., San Diego.

Cantor, C. R. and P. R. Schimmel. 1980. Chapter 13: *X-ray Crystallography.* In: *Biophysical Chemistry.* 1371 pages. W. H. Freeman and Company, New York.

Rhodes, G. 1993. *Crystallography Made Crystal Clear: A Guide for Users of Macromolecular Models.* 202 pages. Academic Press, Inc., San Diego.

12

Appendix 1

Molecular Weights of Commonly Used Chemicals

Chemical	Molecular Weight	Molarity
ACES	182.2	
Acetate (Na salt)	82.0	
Acetic Acid, glacial	60.05	17.4
Acetone	58.1	
Acrylamide	71.1	
Adenosine Triphosphate (ATP, disodium salt)	605.2	
β-Alanine	89.1	
Amido Black 10B	616.5	
Ammonium Hydroxide (NH_4OH)	35.0	14.5 (30%)
Ammonium Persulfate	228.2	
Ammonium Sulfate [$(NH_4)_2SO_4$]	132.1	
Aprotinin	~6500	
Barbital (barbituric acid)	128.1	
Bicinchoninic Acid (BCA)	420.5	
Bicine	163.2	
Bis-Acrylamide (N, N' methylenebis-acrylamide)	154.2	
Boric Acid	61.8	
5-Bromo-4-Chloro-3-Indolyl Phosphate (BCIP)	348.7	
Bromophenol Blue (sodium salt)	692.0	
Cacodylate (sodium salt trihydrate)	214.0	
CAPS	221.3	
CHAPS	614.9	
CHES	207.3	
Chloroform	119.4	
4-Chloro-1-naphthol	178.6	
Citric Acid	192.1	
Coomassie Blue R-250	826.0	
Copper Sulfate ($CuSO_4$)	159.6	

A1

Chemical	Molecular Weight	Molarity
Deoxycholate (DOC)	392.6	
Dimethylformamide (DMF)	73.1	
Dithiothreitol (DTT)	154.3	
Ethanol	46.1	
Ethylene diamine tetraacetic acid (EDTA)	292.2	
Ethylene bis(oxyethylenenitrilo)- tetraacetic acid (EGTA)	380.35	
Ethylene Glycol	62.1	
Formaldehyde	30.0	12.1 (37%)
Formate (sodium salt)	68.0	
Formic Acid	46.0	23.4
Glutaraldehyde	100.1	
Glycerol	92.1	
Glycine	75.1	
Glycine-HCl	111.5	
Glycine-NaOH	97.1	
Glycylglycine	132.1	
Guanidine Hydrochloride	95.5	
HEPES	238.3	
Hydrochloric Acid	36.5	12.1 (36.5 - 38%)
Hydrogen Peroxide (H_2O_2)	34.0	8.8 (30%)
Isopropanol	60.1	
Isopropyl-β-D-thiogalactopyranoside (IPTG)	238.3	
Leupeptin	493.6	
Lithium Dodecyl Sulfate (LiDS)	272.3	
Magnesium Chloride ($MgCl_2$)	95.2	
2-Mercaptoethanol	78.1	14.4
MES	195.2	
Methanol	32.0	
MOPS	209.3	
Nitric Acid	63.0	16 (70%)
p-Nitro Blue Tetrazolium Chloride (NBT)	817.6	
Pepstatin A	686	
Perchloric Acid	100.5	11.6 (70%)
Phenylmethylsulfonyl Fluoride (PMSF)	174.2	
Phosphoric Acid	80.0	18.1 (85%)

A1

Chemical	Molecular Weight	Molarity
PIPES	302.4	
Potassium Chloride (KCl)	74.6	
Potassium Hydroxide (KOH)	56.1	
Riboflavin	376.4	
Silver Nitrate (AgNO$_3$)	169.9	
Sodium Azide (NaN$_3$)	65.0	
Sodium Bicarbonate (NaHCO$_3$)	84.0	
Sodium Carbonate (Na$_2$CO$_3$)	106.0	
Sodium Chloride (NaCl)	58.4	
Sodium Citrate (Na$_3$C$_6$H$_5$O$_7$ [·2H$_2$O])	294.1	
Sodium Dodecyl Sulfate (SDS)	288.4	
Sodium Hydroxide (NaOH)	40.0	
Sodium Phosphate, dibasic (Na$_2$HPO$_4$)	142.0	
Sodium Phosphate, monobasic (NaH$_2$PO$_4$)	120.0	
Sodium Tartrate (Na$_2$C$_4$H$_4$O$_6$ [·2H$_2$O])	230.1	
Succinate (free acid)	118.1	
Succinate (disodium salt)	162.1	
Sucrose	342.3	
Sulfuric Acid	98.1	18 (96%)
N, N, N', N'-Tetramethylethylenediamine (TEMED)	116.2	
TES	229.25	
Trichloroacetic Acid (TCA)	163.4	
Tricine	179.2	
Tris	121.1	
Tween 20	1228	
Urea	60.1	
Zwittergent 3-14	363.6	

A1

Note: An x% solution of compound A contains x grams of compound A
in 100 ml of solvent.

Appendix 2

Molecular Weights and Isoelectric Points of Selected Proteins

Protein	Molecular Weight	IEP
Cytochrome c	11,700	
Ribonuclease	13,700	
Lysozyme	14,300	
*Hemoglobin	15,500	
Myoglobin	17,200	
*β-Lactoglobulin	18,400	
Papain	23,000	8.75
Carbonic Anhydrase	29,000	
Carboxypeptidase	34,600	6.0
Pepsin	35,000	
*Glyceraldehyde-3-Phosphate Dehydrogenase	36,000	
*Lactate Dehydrogenase	36,000	
Tropomyosin	36,000	
*Alcohol Dehydrogenase (Yeast)	37,000	
*Aldolase	40,000	6.1
*Alcohol Dehydrogenase (Liver)	41,000	
*Enolase	41,000	
Ovalbumin	43,000	4.8
*Fumarase	49,000	
Leucine Aminopeptidase	53,000	
*Glutamate Dehydrogenase	53,000	
*Pyruvate Kinase	57,000	
*Catalase	60,000	5.8
Serum Albumin	68,000	
*Phosphorylase a	94,000	
*β-Galactosidase	130,000	
*Myosin Heavy Chain	220,000	

* exists as oligomer in native state

A2

from Practical Handbook of Biochemistry and Molecular Biology. 1989. G.D. Fasman, ed. 601 pages. CRC Press, Inc., Boca Raton, Florida.

Appendix 3

Ammonium Sulfate Precipitation Table

Grams of Ammonium Sulfate to Add to a 1 Liter Solution

Starting Concentration	Final Concentration									
	5%	10%	15%	20%	25%	30%	35%	40%	45%	50%
0%	27	55	84	113	144	176	208	242	277	314
5%		27	56	85	115	146	179	212	246	282
10%			28	57	86	117	149	182	216	251
15%				28	58	88	119	151	185	219
20%					29	59	89	121	154	188
25%						29	60	91	123	157
30%							30	61	92	126
35%								30	62	94
40%									31	63
45%										31

A3

Final Concentration

Starting Concentration	55%	60%	65%	70%	75%	80%	85%	90%	95%	100%
0%	351	390	430	472	516	561	608	657	708	761
5%	319	357	397	439	481	526	572	621	671	723
10%	287	325	364	405	447	491	537	584	634	685
15%	255	292	331	371	413	456	501	548	596	647
20%	223	260	298	337	378	421	465	511	559	609
25%	191	227	265	304	344	386	429	475	522	571
30%	160	195	232	270	309	351	393	438	485	533
35%	128	163	199	236	275	316	358	402	447	495
40%	96	130	166	202	241	281	322	365	410	457
45%	64	97	132	169	206	245	286	329	373	419
50%	32	65	99	135	172	210	250	292	335	381
55%		33	66	101	138	175	215	256	298	343
60%			33	67	103	140	179	219	261	305
65%				34	69	105	143	183	224	266
70%					34	70	107	146	186	228
75%						35	72	110	149	190
80%							36	73	112	152
85%								37	75	114
90%									37	76
95%										38

adapted from Protein Purification: Principles and Practice. 1982. R.K. Scopes. 282 pages. Springer-Verlag, New York.

A3

Appendix 4

Spectrophotometer Linearity

In general, the range of absorbances that a spectrophotometer can read accurately is limited both by stray light that reaches the photomultiplier as well as by the sensitivity of its optics and electronics. 0.1% stray light means that a solution with OD = 3.0 (0.1% transmittance) will appear to have 0.2% transmittance or OD = 2.7. Higher concentrations of absorbing species will never exceed an apparent OD of 3.0, but many instruments will show a deterioration well below this level. In fact, the range of linearity of most mid-range spectrophotometers is below OD = 1.0 - 1.8. Thus, it is important to test your spectrophotometer in order to ascertain its particular range of linearity. To do this, prepare a 0.5% solution of BSA and a series of dilutions to be measured. Plot the OD_{280} versus the concentration. The results will be linear up to the break, after which the OD is always below the extrapolated value (dashed line):

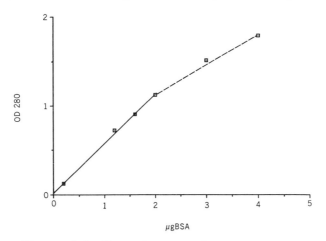

Figure A.1. Test of spectrophotometer linearity.

Recommendations: To be on the safe side, never exceed OD 1 at any wavelength relative to an absorbance by water of 0. Also, keep slit width settings to a minimum, since wider slit widths may also limit the linearity range.

A4

Appendix 5

Suppliers and Addresses

Aldrich Chemical Company, Inc.
1001 West Saint Paul Avenue
Milwaukee, WI 53233
Tel. (414) 273-3850
(800) 558-9160

Amicon
24 Cherry Hill Drive
Danvers, MA 01923
Tel. (617) 777-3622
(800) 343-1397

Baxter HealthCare
1 Baxter Parkway
Deerfield, IL 60015
Tel. (708) 948-2000

Beckman Instruments, Inc.
2500 Harbor Blvd.
Fullerton, CA 92634
Tel. (714) 871-4848
(800) 742-2345

Bethesda Research Laboratories
P.O. Box 6009
Gaithersburg, MD 20877
Tel. (301) 840-8000
(800) 638-8992

Bio-Rad Laboratories
1414 Harbour Way, S.
Richmond, CA 94804
Tel. (415) 232-7000
(800) 4-BIORAD

BioSepra Inc.
140 Locke Drive
Marlborough, MA 01752
Tel. (508) 481-6802
(800) 752-5277

Boehringer-Mannheim Corp.
9115 Hague Road
P.O. Box 50414
Indianapolis, IN 46250-0414
Tel. (800) 262-1640

B. Braun Instruments
824 Twelfth Ave.
Bethlehem, PA 18018
Tel. (215) 868-0300
(800) 258-9000

BDH Chemicals, Ltd.
Broom Road
Poole, Dorset
BH12 4NN UK

A5

Calbiochem-Novabiochem
Corporation
P.O. Box 12087
La Jolla, CA 92039-2087
Tel. (619) 450-9600
(800) 854-3417

CUNO, Inc.
400 Research Pkwy.
Meriden, CT 06450
Tel. (203) 237-5541
(800) 231-2259

Eastman Kodak Co.
343 State St., Building 701
Rochester, NY 14650
Tel. (716) 458-4014
(800) 225-5352

Fisher Scientific
50 Fadem Road
Springfield, NJ 07081
Tel. (201) 467-6400
(800) 766-7000

Fluka Chem. Corp.
980 S. Second St.
Ronkonkoma, NY 11779
Tel. (516) 467-0980
(800) FLUKA-US

Gaulin Corporation
44 Garden Street
Everett, MA 02149

Hamilton Bonaduz AG
case postale 26
CH-7402 Bonaduz
Switzerland
Tel. 041 081 37 14 33

Hamilton Instruments
P.O. Box 100030
Reno, NV 89520

Hampton Research
25431 Cabot Road, Suite 205
Laguna Hills, CA 92653-5527
Tel. (714) 699-1040
Fax (714) 586-1453
E-mail: XTALROX@AOL.COM

Hoefer Scientific Instruments
P.O. Box 77387
654 Minnesota St.
San Francisco, CA 94107
Tel. (415) 282-2307
(800) 227-4750

Hunt Mfg. Co.
(Speedball India Ink)
Statesville, NC

ICN Pharmaceuticals, Inc.
3300 Hyland Avenue
Costa Mesa, CA 92626
Tel. (714) 545-0113
(800) 854-0530

Integrated Separation Systems
21 Strathmore Road
Natick, MA 01760
Tel. (508) 655-1500
(800) 433-6433

International Products
P. O. Box 70
Burlington, NJ 08016
Tel. (609) 386-8770

Kontron Instruments Inc.
  (Arrow International)
9 Plymouth St.
Everett, MA 02149
Tel. (617) 389-6400
(800) 343-3297

E .M. Merck
5 Skyline Drive
Hawthorne, NY 10532
Tel. (914) 592-4660
(800) 831-3662

Millipore Corp.
80 Ashby Rd.
Bedford, MA 01730
Tel. (617) 275-9200
(800) 645-5476
(800) 632-2708

New Brunswick Scientific Co., Inc.
44 Talmadge Road
Edison, NJ 08818
Tel. (201) 287-1200
(800) 631-5417

New England Nuclear
549 Albany Street
Boston, MA 02118
Tel. (617) 350-9153
(800) 551-2121

Pellikan AG (Pelikan India Ink)
D-3000
Hanover 1
Germany

A5

Pharmacia Biotech
S-75182
Uppsala, Sweden
Tel. 011 46 18 16 5000

800 Centennial Ave.
P.O. Box 1327
Piscataway, NJ 08855-1327
Tel. (201) 457-8000
(800) 526-3593

Pierce Chemical Co.
P.O. Box 117
Rockford, IL 61105
Tel. (815) 968-0747
(800) 8-PIERCE

Polysciences, Inc.
400 Valley Rd.
Warrington, PA 18976
Tel. (215) 343-6484
(800) 523-2575

Sartorius Filters, Inc.
30940 San Clemente St.
Hayward, CA 94544
Tel. (415) 487-8220
(800) 227-2842

Schleicher & Schuell, Inc.
10 Optical Ave.
Keene, NH 03431
Tel. (603) 352-3810
(800) 245-4024

Serva Fine Biochemicals Inc.
200 Shames Drive
Westbury, NY 11590
Tel. (516) 333-1575
(800) 645-3412

Sigma Chemical Co.
P.O. Box 14508
St. Louis, MO 63178
Tel. (314) 771-5750
(800) 325-3010

Spectrum Medical Industries, Inc.
60916 Terminal Annex
Los Angeles, CA 90054
Tel. (213) 650-2100
(800) 634-3300

Thomas Scientific
99 High Hill Road
P.O. Box 99
Swedesboro, NJ 08085
Tel. (609) 467-2000
(800) 524-0027

Vector Laboratories, Inc.
30 Ingold Road
Burlingame, CA 94010
(415) 697-3600
(800) 227-6666

VWR Scientific
P.O. Box 7900
San Francisco, CA 94120
Tel. (415) 467-6202
(800) 257-8407

Whatman, Inc.
9 Bridewell Pl.
Clifton, NJ 07014
Tel. (201) 773-5800
(800) 631-7290

A5

INDEX

Acetone, 67, 77, 94, 95, 357, 387; precipitation, 86

N-Acetyl glucosamine, 331, 332

Acrylamide, 78, 109-111, 114, 120, 124, 133, 135-137, 139-143, 147, 149, 156-158, 160, 169, 170, 176, 179, 187, 205, 232, 387. *See also* Bis-acrylamide, Polyacrylamide, SDS-PAGE

ADA, 7, 9, 235

Adenosine 5'-triphosphate (ATP), 11, 20, 51, 78, 329, 387

Adsorb, adsorption, 232, 233, 240, 246, 261, 264, 265, 292, 293, 328, 344, 346

Affinity chromatography, *see* Chromatography, affinity

Agarose, 132, 137, 179, 184, 187, 191, 232, 246, 248, 277, 278, 303, 304, 335, 338

Aggregate, aggregation, 15, 44, 50, 89, 91, 93, 110, 123, 156, 180, 184, 211, 214, 354, 362, 363, 369, 374, 375, 379, 382

Albumin, bovine serum (BSA), 20, 59, 62, 64-67, 90, 97, 207, 208, 210-212, 391, 395

Alkaline phosphatase, 9, 20, 171, 196, 197, 200, 207, 209, 213, 222

Alumina, 38

Amido Black, 200, 216, 387

Amino acid analysis, 198, 221

Ammonium persulfate, 110-114, 116, 120, 137, 139, 140, 142-144, 158-162, 176, 178-179, 183, 387

Ammonium sulfate, 14, 21, 76, 96, 264, 294, 314, 327, 345, 357,

363, 370, 387; precipitation, 91-93, 392, 393

Ampholytes, 78, 133, 176-180, 182-187, 191, 192

Anion exchange, exchanger, 4, 103, 232, 234, 235, 236, 237, 246, 259, 264. *See also* Chromatography, ion exchange

Antibody, 43, 46, 87-90, 108, 196-198, 207-216, 218, 221, 275, 300; column, *see* Chromatography, immunoaffinity; purification of, 311-318

Aprotinin, 24, 387

ATP, *see* Adenosine 5'-triphosphate

Autoradiography, 90, 134, 138, 145, 198, 218

Azide, 8, 10, 212, 259, 290, 306, 317, 323, 334, 368, 389

Bacillus subtilis, 35, 38

Bacteria, bacterial, 10, 29, 31, 36, 43, 44, 46, 47, 50, 87, 149, 312, 363, 368

Batch chromatography, *see* Chromatography, batch

BCA protein assay, *see* Bicinchoninic protein assay

Bicinchoninic (BCA) protein assay, 9, 72-74, 76-79, 387

Bis-acrylamide, 110, 111, 137, 140, 143, 156-158, 176, 278. *See also* Acrylamide, Polyacrylamide, SDS-PAGE

Blender, 32, 33, 35

Blocking membranes, 197, 200, 207, 210, 212
BLOTTO, 210
Blue dextran, 285, 292; *see also* Dextran
Borate, 7-9, 334, 357
Bradford protein assay, 9, 60, 62-67, 76-79
Brain, 33, 35
Bromophenol blue, 110, 112, 122, 124, 146, 157, 159, 176, 180, 191, 387
BSA, *see* Albumin, bovine serum
Buffer, 2-10; coupling, 305, 306; crystallization, 362-364, 377-378; electrode (electrophoresis), 111, 117, 122, 125, 159, 161, 162, 164, 190; equilibration (IEF), 189; erasing, 200, 218; Good's, 4, 9, 363, 377; ion exchange chromatography, 234-238; loading (IEF), 180; permeability, 3; sample (PAGE), 89, 90, 112, 121, 123, 136, 138, 144, 159, 161, 164, 165, 183, 373; separating, 158; stacking, 158; transfer, 196, 201, 202, 205, 206, 212
Buffer exchange, 100, 294-295, 345

Cacodylate, 7, 9, 76, 387
Calcium, 11, 221, 341
Carbohydrate, 304, 331
Carbonate, 7, 9, 72, 389
Cathodic drift, 185, 186
Cation, divalent, 11, 12, 14, 86, 95, 329, 333, 335, 337
Cation exchange, exchanger, 4, 232-238, 246, 259. *See also* Chromatography, ion exchange

Cell suspensions, 31
Cellulose, 232, 248
Chaotropic, 307, 309, 339, 343, 346
Chaperones, 51
CHAPS, 16, 17, 76, 187, 387
Cheesecloth, 30
Chelating, chelators, 8, 11, 19, 77, 93, 341
Chloronaphthol, 200, 209, 212
Cholate, *see* Sodium cholate
Chromatography, affinity, 299-346; batch, 240, 249, 264, 265, 333; dye ligand, 336-340; gel filtration, 4, 13, 19, 96, 99, 227, 228, 253, 279-295, 314, 317, 322, 323, 327, 363; hydrophobic interaction (HIC), 309, 339, 344-346; hydroxyapatite, 4, 246; immobilized metal affinity (IMAC), 341-343; immunoaffinity (immunopurification), 314, 320-323; ion exchange, 16, 96, 103, 227, 228, 230-265, 270, 273, 281, 294, 305, 310, 314, 327, 330, 335, 345; lectin affinity, 300, 305, 330, 331-335; nucleic acid affinity, 324-330. *See also* anion exchange, cation exchange
Chromatofocusing, 234
Cibacron Blue F3GA, 336-338
Citrate, 7, 9, 68, 76, 90, 341, 357, 364, 389
CM (cation exchanger), 232, 237, 238, 246, 248. *See also* Chromatography, ion exchange
CNBr, *see* Cyanogen bromide
Cofactor, 11, 46, 50, 149, 227, 262, 263, 273, 275, 293, 300, 324, 329, 346, 372, 381

Column, capacity, 247, 306, 316, 321, 322, 332, 342, 343, 345; dead space, 243, 252, 281, 292; dimensions, 250, 274, 342; flow rate, 252, 260, 262, 272, 277-279, 281, 284, 287, 291, 292, 294, 303, 315, 322, 333, 334; packing, 249-252, 261, 262, 282-285, 292, 305, 333; regeneration, 247, 258, 259, 261-263, 289-291, 323, 328, 334, 338, 340, 343, 345, 346; reservoir, 243, 244, 250, 251, 254, 279, 281, 284, 287;

Comb, 114, 116-118, 123, 162, 163, 168, 179, 189

Concanavalin A (ConA), 210, 331, 332

Contamination, microbial, 10, 36, 100, 321, 323, 334, 363, 368

Coomassie, gel destain, 75, 127, 167, 168, 176, 182; gel stain, 75, 127, 128, 157, 167, 168, 176, 182, 385; gel staining, 109, 123, 126-128, 133, 138, 148, 167, 171, 175, 180, 182, 187, 190, 221

Copper, elution with, 341

Counter-ion, 230, 232-235, 239, 240, 261

Coupling, affinity ligand, 301, 303, 304, 305-307, 326, 327, 329, 333

Coupling buffer, see Buffer, coupling

Coverslip, crystallization, 352, 359-361, 364-368

Cracking, gel   120, 135

Critical micelle concentration (CMC)   13-15, 17, 19

Crude extract   30, 42, 87, 103, 190, 196

Crush test, 372

Crystal, 227, 356, 358, 359, 367-375, 377, 378, 380-382. See also, Crystallization

Crystal Screen kit, 364, 365, 375, 376, 382, 383

Crystallization, 49, 93, 96, 227, 228, 352-382

Crystallization buffer, see Buffer, crystallization

Cultured cells   31, 33, 42, 89

Current   117, 122, 124, 125, 166, 181, 196, 202, 205

Cuvettes   58-62, 64, 67, 68, 72

Cyanogen bromide (CNBr), 301, 303-305, 315, 324, 326, 330, 333, 335

Dead space, column, see Column, dead space

DEAE, 232, 233, 235, 237-239, 246, 248, 258, 264, 314. See also Chromatography, ion exchange

Degassing, 120, 251, 260, 282, 291

Denaturant (denaturing agent), 19, 46, 49, 76, 149, 346. See also denaturation, denature

Denaturation, 13, 16, 20, 21, 34, 43, 46, 84, 93, 96, 97, 156, 294, 321, 344. See also denaturant, denature

Denature, 4, 13-15, 46, 49, 62, 68, 72, 84, 94, 95, 107, 123, 149, 156, 157, 166-169, 171, 178, 180, 185-187, 196, 219, 237, 273, 288, 292, 309, 310, 319, 339, 346, 356, 363, 369. See also denaturant, denaturation.

Denaturing agent, see denaturant

Densitometry, 135, 138, 148

Deoxycholate (DOC), see Sodium deoxycholate

Deoxyribonuclease (DNAse), 329

Desalting, 92, 97, 273, 294

Detergent, 13-19, 46, 67, 71, 74, 76, 89, 136, 137, 149, 175, 184, 187, 220, 258, 260, 261, 289, 291, 292, 307, 309, 328, 332, 335, 343, 346, 381; lysis, 31, 42

Dextran, 248, 277, 278. *See also* Blue dextran

Dialysis, dialyze, 13, 49, 92, 104, 294, 308, 317, 322, 323, 333, 334, 338, 342, 363; method, 100-102

Dialysis membrane, *see* Membrane, dialysis

Dielectric constant, 356, 377

Disulfide exchange, 45, 46, 49

Dithiothreitol (DTT), 5, 12, 19, 77, 110, 161, 310, 388

DNA, 78, 93, 220, 324-326, 327, 329, 330

DNAse, *see* Deoxyribonuclease

DNAse footprinting, 329

DOC, *see* Sodium deoxycholate

DOC-TCA precipitation, 70, 74, 85

Dounce homogenizer, 31, 33

Drying a gel, 135

DTT, *see* Dithiothreitol

Dye, 62, 64, 67, 71, 110, 121, 124, 128, 138, 146-148, 161, 166, 169, 186, 197, 216, 217, 285, 300, 373. *See also* Chromatography, dye ligand

Dye ligand chromatography, *see* Chromatography, dye ligand

EDTA, 5, 11, 19, 23, 24, 41, 47, 71, 77, 87, 93, 100, 213, 220, 289, 327, 328, 329, 339, 341, 343, 388

EGTA, 11, 19, 77, 341, 388

Electrode buffer, *see* Buffer, electrode

Electroelution, 149, 171

Electrophoresis buffer, *see* Buffer, electrode

Electrostatic forces, 93, 233, 309, 336, 356

ELISA, 15, 308, 318, 319

Elution, ampholytes, 191; antibodies, 317-319; batch ion exchange, 265; dye ligand columns, 338, 339; *from electrophoresis gels, see* electroelution; gel filtration columns, 281, 285, 286-288, 291-293, 295; hydrophobic interaction columns, 345, 346; immobilized metal columns, 343; immunoaffinity columns, 320, 321-323; ion exchange columns, 232-234, 239-241, 244, 247, 255-257, 261-263; lectins, 332-335; *onto* nitrocellulose, 201-206; nucleic acid affinity column, 327, 329;

Elution profile, 58, 228

Enzymatic lysis, 31, 41

Equilibration, antibody purification columns, 315, 317; dye ligand affinity columns, 338; gel filtration columns, 273, 284, 294, 295; gels, 175, 188, 189, 191; immobilized metal columns, 342; immunoaffinity columns, 322; ion exchange columns, 246, 249, 250, 252, 261-265; lectin affinity columns, 333; nucleic acid columns, 327, 328

Equilibration buffer, *see* Buffer, equilibration

Erasing buffer, *see* Buffer, erasing

Erasing immunoblots, 200, 218

Erythrocytes, 31

Escherichia coli, 35, 37, 38, 41, 47, 51, 136, 190, 327
Ethanolamine, 235, 306, 326
Ethylene glycol, 77, 101, 309, 335, 339, 346, 388

Filter binding assay, 75, 79
Fines, 249, 260, 262, 282, 291
Flow adaptor, 243, 252, 254, 283
Flow rate, column, *see* Column, flow rate
Fluorography, 134, 145
Folding intermediates, 45, 50, 51
Formaldehyde, 130, 139, 388
Fraction collector, 243, 257, 262, 266, 279, 281, 292, 296
Freeze-thaw lysis, 31, 43
Freezing, 21, 22, 123, 211
French press, 31, 33, 36-38, 52
Fusion proteins, 51

Gel drying, *see* Drying a gel
Gel electrophoresis, *see* SDS-PAGE, Nondenaturing gel electrophoresis
Gel filtration, *see* Chromatography, gel filtration
Gel retardation, *see* Chromatography, gel filtration
Gel slicing, 145
Gel swelling, 205, 206
Gelatin, 210
Glass beads, 31, 33, 39, 40, 52
Glass wool, 30, 244
Glutaraldehyde, 134, 185, 206, 304, 388
Glutathione, 12, 48, 49, 52
Glycerol, 9, 19, 22, 42, 77, 98, 110, 112, 123, 133, 135, 149, 157, 159, 183, 188, 189, 388

Glycine, 7, 76, 110, 111, 133, 157, 159, 200, 201, 309, 315, 317, 322, 323, 343, 388
Glycoproteins, 136, 331, 335
Glycyl betaine, 51
Gold, *see* Immunogold
Good's buffers, *see* Buffer, Good
Gradient gels, 108, 140-142
Gradient maker, 141, 142, 233, 244, 257, 266
Grid screen, 374, 375, 376, 382
Grinding tissue, 31, 38
GTP, see Guanosine 5'-triphosphate
Guanidine HCl, *see* Guanidine hydrochloride
Guanidine hydrochloride, 19, 46, 49, 52, 76, 90, 309, 340, 388
Guanosine 5'-triphosphate (GTP), 221, 381

Hanging drop method, 227, 352-382
Heart, 33
HEPES, 7-9, 76, 364, 375, 388
HIC, *see* Chromatography, hydrophobic interaction
Histones, 136, 191, 219, 221
Homogenate, 30, 33
Homogenization, 31-33
Homology search, 381
Horseradish peroxidase, 196, 197, 200, 207, 209, 212, 213, 216, 222
Hydrophobic interaction chromatography (HIC), *see* Chromatography, hydrophobic interaction
Hydroxyapatite, *see* Chromatography, hydroxyapatite
Hydroxysuccinimide, 304, 305

[125]I labeling, 134, 196, 197, 207, 215
IEF, *see* Isoelectric focusing
IMAC, *see* Chromatography, immobilized metal affinity
Immobilized metal affinity chromatography (IMAC), *see* Chromatography, immobilized metal affinity
Immunoaffinity chromatography, *see* Chromatography, immunoaffinity
Immunoblotting, 15, 90, 108, 196-221, 275, 311, 318, 319
Immunodetection, 207, 210, 211, 213-215, 217
Immunoglobulin, 87, 88, 90, 197, 210, 312, 313
Immunogold labeling, 196, 197, 200, 207, 214, 216, 217
Immunoprecipitation, 87-90, 185, 196, 319, 321
Immunopurification, *see* Chromatography, immunoaffinity
Inclusion bodies, 29, 44-51, 149
India ink, 186, 200, 216, 217, 222
Inhibitors, protease, 19, 23, 24
Ion exchange chromatography   see Chromatography, ion exchange
Ionic strength   3, 4, 11, 13, 43, 94, 143156, 168, 205, 231, 234, 235, 253, 255, 257, 258, 260, 261, 263-265, 270, 273, 275, 286, 292, 294, 309, 310, 316, 340
Isoelectric focusing (IEF), 16, 89, 174-191, 219, 231, 237, 362
Isoelectric point   93, 96, 147, 174, 177, 184, 186, 196, 206, 219, 230, 231, 234, 236, 237, 240, 261, 264, 346, 378, 391

Klebsiella pneumoniae 35

Lectin, 210, 220, 300, 301, 305, 330; purification, *see* Chromatography, lectin affinity
Lectin affinity chromatography, *see* Chromatography, lectin affinity
Leupeptin, 19, 24, 388
Ligand, 2, 21, 46, 50, 198, 220, 228, 273, 300, 352, 372, 374, 381; *for* affinity chromatography, 300-346
Lipid, 15, 71, 74, 78, 175, 186, 258, 289
Lithium dodecyl sulfate, 14, 17, 120, 388
Liver, 33
Loading buffer, *see* Buffer, loading
Lowry protein assay, 9, 68, 70-72, 74, 76-78, 79
Lyophilization, 22, 41, 103
Lysis, 29-43
Lysozyme, 41, 47, 129, 219, 355, 391

Magnesium, 11, 95, 329
Membrane, cell, 3, 9, 18, 19, 30, 35, 48; dialysis, 100, 102, 104; filtration, 363; immunoblotting, 196, 197, 199, 201-203, 205-211, 214-216, 218-221; lipid, 15; ultrafiltration, 97-99, 363
Membrane protein, 2, 13-15, 18, 19, 67, 143, 184, 187, 331, 332, 335, 344
2-Mercaptoethanol, 12, 19, 77, 90, 110, 112, 136, 176, 180, 188, 189, 191, 200, 218, 310, 318, 388

Metal ions, 3, 5, 8-11, 93, 134, 259, 339, 346, 368, 381. *See also* Chromatography, immobilized metal affinity

2-Methyl-4-pentanediol (MPD), 356, 357, 383

Micelle, 13, 14, 17, 136

Microcrystal, 354

Microscope, phase contrast, 33, 35, 49

Modification, protein, 2, 50, 108, 171, 175, 184, 362, 382

Molecular weight, *of* common chemicals, 387-389; determination, 14, 108, 136, 146, 147, 169, 170, 272, 273, 275, 288; *and* electrophoretic separation, 108, 136, 140, 143, 174, 184, 185, 187, 188; *and* immunoblotting, 197, 205

Molecular weight markers, gel filtration, 288; immunoblotting, 205, 209, 211, 215; SDS-PAGE, 122, 129, 138, 146, 167, 169, 170; 2-D gels, 191

MOPS, 4, 7-9, 235, 388

MPD, *see* 2-methyl-4-pentanediol

Muscle, 33

NAD, *see* Nicotine adenine dinucleotide

Native gel electrophoresis, *see* Nondenaturing gel electrophoresis

Native structure, 46, 48, 50

NEPHGE, *see* Nonequilibrium pH gradient electrophoresis

Net charge, 136, 174, 230-234, 236, 240

Neurospora, 38

Nickel, IMAC elution, 341

Nicotine adenine dinucleotide (NAD), 78, 324, 336, 339

Nitrocellulose membrane, 197, 199, 201-204, 206-211, 213-222

Nondenaturing gel electrophoresis, 156-171, 196, 219

Nonequilibrium pH gradient electro-phoresis (NEPHGE), 185, 186

Nonidet P-40, 76, 184, 187, 289, 335

Nucleation, 354

Nucleic acids, 9, 43, 58, 60, 61, 134, 138, 184, 289, 300, 301, 305. *See also* Chromatography, nucleic acid affinity

Nucleic acid affinity chromatography, *see* Chromatography, nucleic acid affinity

Nylon membrane   197, 210, 216, 218, 222

Octyl glucoside, 13, 15, 17, 76, 333

Octyl sepharose, 344, 346

Oligonucleotide, 324-326, 329, 330

Oligosaccharides, 9, 331

Organic precipitants, 84, 86, 94, 95, 356, 357, 368

Organic solvent, 13, 84, 86, 102, 289, 377; lysis, 31, 43; precipitation, 94-95

Organosilane, 359, 360

Osmolarity, 352, 372, 379

Osmotic shock, 31, 43, 102

Osmotic stress, 51

Oxidation, 12, 184, 362

Pansorbin cells, 88

Paracoccus, 37

Parafilm, 40, 111, 148, 158, 189
PEG, *see* Poly(ethylene glycol)
Pepstatin A, 19, 24, 388
Percent formulas (%C, %T), 137, 161, 178, 183
Peristaltic pump, 21, 140-142, 243, 244, 266, 277, 279, 281, 287, 296
pH, *and* affinity column elution, 309, 317; *and* antibody elution, 312, 315, 317, 319; *and* buffer concentration 8; *and* buffers, 3-10, 377-378; *and* CMC, 13; *and* colloidal gold staining, 217; *and* coupling, affinity reagent, 305-307; *and* crystallization, 354, 355, 359, 364, 374, 375, 377-379; *and* dye ligand elution, 337, 339, 340; *and* freezing proteins, 21; *and* gel filtration, 275, 289; gradient, IEF, 182, 185, 186; gradient, ion exchange chromatography, *see* Chromatofocusing; *and* HIC elution, 346; *and* IMAC elution, 343; *and* immunoaffinity elution, 322, 323; *and* immunoprecipitation, 89, 90; *and* IEF, *see* Isoelectric point, Net charge; *and* ion exchange chromatography, 231-234, 236-238, 240, 246, 247, 250, 252, 253, 261-264; *and* membrane protein association, 19; *and* nondenaturing gel electrophoresis, 156, 157, 161; *and* nucleic acid affinity elution, 329; *and* protease inhibitors, 22, 23; *and* protein refolding, 49; *and* protein solubility, 91, 95, 354, 355,

374, 375, 377; range, IEF, 177, 178, 180, 183, 185, 187
Phenyl Sepharose, 344, 345, 346
Phenylmethylsulfonyl fluoride (PMSF), 19, 23, 24, 388
Phosphate, 4, 6-10, 19, 76, 335, 339, 357, 370, 389; determination, 326
Phosphocellulose, 330
pK, 3, 4, 7, 9, 14, 235, 364, 377
Plant tissue, 31, 38
PMSF, *see* Phenylmethylsulfonyl fluoride
Polyacrylamide, gels, 108, 120, 128, 148, 156, 169, 171, 174, 187-191, 196, 201-203, 207, 220, 275, 318; matrix, 277, 278, 303, 304
Poly(ethylene glycol) (PEG), 100, 101, 132, 356, 368, 383; precipitation, 96
Polyethyleneimine, 329
Poly(vinylidene difluoride) (PVDF) membrane, 221
Ponceau S, 216, 217, 222
Potter-Elvehjem homogenizer, 33
Power supply, 109, 124, 150, 166, 175, 188, 192, 199, 202, 222
Precipitant, *see* Precipitating agent
Precipitate, protein, 21, 44, 67, 68, 352, 354, 355, 358, 363, 369, 370, 374, 375, 378, 380
Precipitating agent, 49, 352, 355-359, 362, 365, 366, 368-370, 374, 377-380, 382
Precipitation, 44, 45, 70, 71, 74, 84, 354, 370, 375, 377, 380, 382, 392, 393. *See also* Acetone precipitation, Ammonium sulfate precipitation, Immunoprecipitation, Organic solvent precipitation, PEG

precipitation, TCA
  precipitation
Precrystalline aggregates, 369, 374,
  375
Preferential hydration, 51
Preparative isoelectric focusing,
  191
Prestained molecular weight
  markers, 138, 147, 205, 211
Prosthetic group, 50
Protease inhibitors, 19, 23, 24, 47,
  293
Protease, 11, 19, 23, 24, 32, 89,
  230, 262, 289, 321
Protein A, 87, 88, 90, 104, 210,
  215, 312, 313, 315, 318, 320
Protein G, 90, 104, 312, 313, 315,
  318
Proteolysis, 2, 23, 43, 108, 136,
  138, 156, 293, 323, 362
Pseudomonas, 37
PVDF, *see* poly(vinylidene
  difluoride)

Radiolabeled proteins, 90, 134,
  145, 215, 217
Recorder, chart, 244, 257, 266,
  279, 281, 296
Reducing agents, 12, 19, 77, 123,
  149, 161, 275
Reductive methylation, 382
Regeneration, column, 247, 258,
  261-263, 289, 291, 323, 328,
  334, 338, 340, 343, 345, 346
Renaturation, protein, 13, 45, 46,
  48-50, 149, 198, 220
Riboflavin, 137, 142, 161, 389
R_f, 146, 147, 169, 170
RNA, 78, 93, 220, 324, 371

Saccharomyces cerevisiae, *see*
  Yeast

Salt, *and* antibody stability, 318;
  and crystallization, 355-357,
  368, 370, 372, 373; *and* dye
  ligand chromatography, 336,
  337, 339; *and* electrophoresis,
  125, 165, 184; *and* gel
  filtration, 275, 286, 292-294;
  and HIC, 344, 346; *and*
  IMAC, 341; *and*
  immunopurification, 323; *and*
  ion exchange chromatography,
  232, 233, 238-241, 253, 256,
  258, 262; *and* lectin affinity
  chromatography, 334, 335;
  and membrane proteins, 19;
  and protein stability, 11, 86.
  See also Ammonium sulfate
  precipitation
Salting out, *see* Ammonium sulfate
  precipitation
Sample buffer, *see* Buffer, sample
Sand, grinding, 38
Sand bath, 109, 150
Saturation, *and* affinity column
  elution, 308; *of* ammonium
  sulfate solubility, *see*
  Ammonium sulfate saturation;
  of antibody binding, 211; *of*
  buffer solubility, 8; *of* protein
  solubility, 352, 354, 355, 380
SDS, *see* Sodium dodecyl sulfate
SDS-PAGE, 156, 169, 171, 174,
  188-192, 201, 211, 220, 275,
  318, 362, 372, 373; methods,
  108-149. *See also*
  Acrylamide, Bis-acrylamide,
  Isoelectric focusing,
  Nondenaturing gel
  electrophoresis,
  Polyacrylamide, SDS-urea
  PAGE, Two-dimensional gel
  electrophoresis
SDS-urea PAGE, 136, 143, 144,
  186, 219

Sealing gel plates, 137, 179
Sephadex, *for* gel filtration, 278, 289, 291, 294, 363; *for* ion exchange, 247, 248, 252, 258, 260; *for* protein concentration, 100-104
Sepharose, affinity columns, 301, 305, 306, 315, 324-327, 333, 335, 340-342, 344-346; antibody adsorption columns, 87, 88, 90, 315; gel filtration, 278, 289, 290 ion exchange, 237-239, 246, 248, 252, 258, 264
Sequencing, protein, 176, 198, 221
Silver nitrate   130, 139, 389
Silver staining   108, 109, 123, 126, 128, 150, 167, 191; procedure, 129-134
Size standards, *see* Molecular weight markers
Slurry, chromatography matrix, 250-252, 265, 282, 283, 305; water-ice, 92
Smiling, gel, 125
Sodium cholate, 14, 17, 76
Sodium deoxycholate (DOC), 14, 17, 76, 309, 388; pH and solubility, 14. *See also* DOC-TCA precipitation
Sodium dodecyl sulfate (SDS), 14, 17, 389; *and* antigenicity, 211; *in* electrophoresis, *see* SDS-PAGE, SDS-urea PAGE; interference, 71, 76, 85; removal, 149, 187, 205; *and* transfer efficiency, 206
Solubility, buffers, 3, 9; detergents, 13; proteins, 45, 50, 51, 84, 86, 93-96, 227, 352, 354, 355, 358, 379, 380, 382; salts, 91
Solubilization, crystals, 373; immunoprecipitates, 90; *from* inclusion bodies, 44-51;

membrane proteins, 13-15, 17-19; SDS, 123
Sonication, 18, 31, 34, 35, 47
Sorbitol, 51
Sparse matrix, 49, 359, 364, 369, 374, 378, 382
Spectrophotometer, 58, 59, 60, 62, 68, 72, 148, 244, 395
Standard curve, 63-67, 69, 70, 73, 74, 146-148, 288
Staphylococcus aureus 87-90, 104, 312
Streptomycin, 329
Stoichiometry, 138
Storage of protein, 10, 12, 21, 22, 30
Stray light, 60, 395
Streaking, 123, 175, 180, 184, 191, 285
Substrate, *and* affinity chromatography, 300, 308, 321, 346; *and* crystallization, 381; *for* immunoblot development, 212, 213; *and* renaturation, 50, 149;
Sucrose, 71, 77, 140, 141, 219, 389
Supersaturation, 354
Swelling, chromatography matrix, 87, 247, 249, 258, 260, 282, 326

TBS buffer, 42, 200, 206-214, 216, 218, 219
TCA, *see* Trichloroacetic acid
TE buffer, 41, 289
TEMED, 389; *in* IEF, 178, 179, 183; *in* Nondenaturing gel electrophoresis, 159, 160, 162; *in* SDS-PAGE, 110, 112-114, 116, 120, 139-144
Temperature, *and* chromatography matrix swelling, 249, 282; *and* CMC, 13; *and* crystallization,

367, 368, 379; *and* detergent solubility, 14, 15, 120, 123; *and* lysozyme lysis, 41; *and* pK variability, 3-6, 9; *and* protein migration on gels, 124; *and* protein renaturation, 46, 50, 51; *and* protein solubility, 19, 91, 94, 354, 355; *and* protein stability, 20, 21

Titer, 318

Transfer buffer, *see* Buffer, transfer

Trichloroacetic acid (TCA), 78, 104, 389; *in* filter binding assay, 75; *for* fixing gels, 132, 137, 185, 187; *in* IEF, 176, 182, 185, 187; *in* immunoblotting, 216; precipitation, 84-85. *See also* DOC-TCA precipitation

Tris, 7, 8, 9, 363, 389; interference, 76; ion exchange chromatography, 235; metal chelation, 11; pK variability, 5, 8

Triton X-100, 15, 17, 42, 48, 52, 71, 76, 87-89, 104, 133, 136, 176, 180, 184, 187, 328, 335, 346

Triton X-114, 15, 17, 19

Tube gels, 108, 149, 174, 188

Tubing, chromatography, 243, 244, 260, 261, 266, 279, 281, 287, 288, 291, 296; dialysis, 100-102; gradient gel making, 140, 142

Tween, 15, 17, 76, 200, 210, 212, 214, 216, 346, 389

Two-dimensional gel electrophoresis, 174, 188-191

Ultracentrifugation, 89, 97, 253

Ultrafiltration, 10, 92, 97, 99, 253, 294, 317, 363

Ultra-thin gels, 187

Urea, 19, 46-49, 76, 100, 133, 136, 143, 144, 175, 176, 178-181, 184, 186, 187, 190, 219, 273, 309, 339, 340, 343, 346, 389

UV detector, 244, 266, 279, 296

Void volume, 285

Water, purity, 10, 362

Western blotting, *see* Immunoblotting

X-ray crystallography, 352, 373

Yeast, 31, 36-38, 40, 41, 209, 371

Zinc, 11, 221, 341

Zwittergent, 16, 17, 187, 389